MAPS of the WORLD'S OCEANS

An ILLUSTRATED CHILDREN'S ATLAS to the SEAS and All the CREATURES and PLANTS that LIVE THERE

MAPS of the WORLD'S OCEANS

An ILLUSTRATED CHILDREN'S ATLAS to the SEAS and All the CREATURES and PLANTS that LIVE THERE

ENRICO LAVAGNO
and ANGELO MOJETTA

Illustrated by
SACCO VALLARINO

BLACK DOG
& LEVENTHAL
PUBLISHERS
NEW YORK

Contents

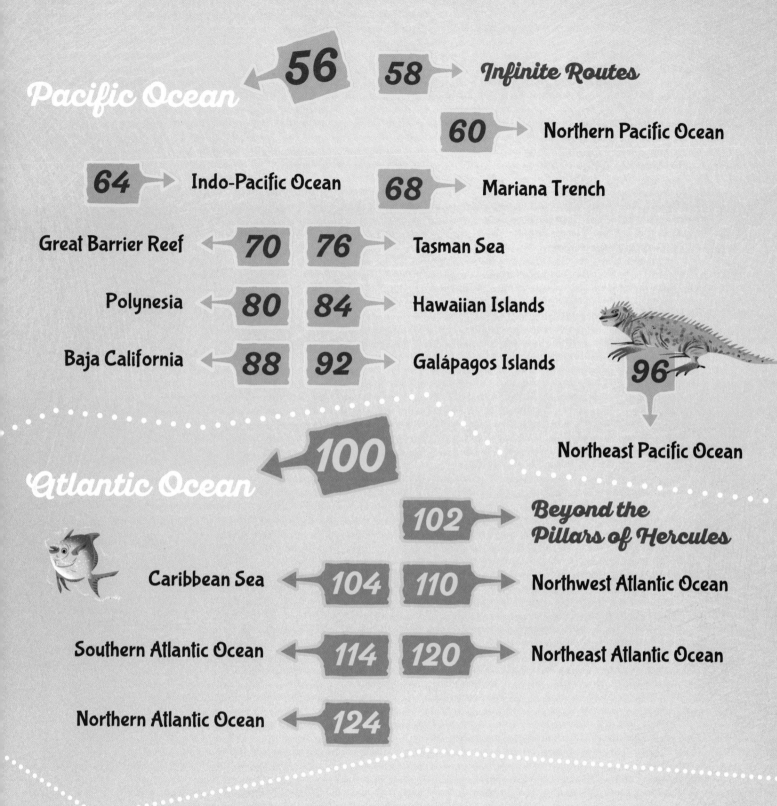

Emergence of Continents and Oceans

600 million years ago—Proterozoic

If we were to go back 600 million years, we would not recognize our planet. It had one huge ocean called *Panthalassa*, or "all the seas," which was already full of life that had been developing for millions of years.

300 million years ago—the Carboniferous

Over the next 300 million years (which for the Earth is more or less like two human years) the planet went through many changes. The continents shifted and a new ocean called the Paleo-Tethys begins to form.

200 million years ago—Jurassic Pangea

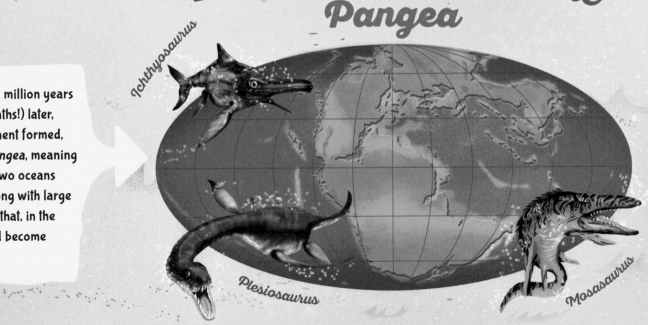

One hundred million years (or eight months!) later, a supercontinent formed, known as *Pangea*, meaning "all lands." Two oceans remained along with large interior seas that, in the future, would become new oceans.

Ichthyosaurus

Plesiosaurus

Mosasaurus

150 million years ago—Cretaceous

Tyrannosaurus rex

Hesperornis

Dakosaurus

Tylosaurus

During the dinosaur era, Pangea split apart and its fragments went in different directions. The interior seas expanded and became deeper. Earth was still very different from the way it is today, but certain familiar things began to appear, such as the Pacific and Atlantic oceans, Africa, and South America.

Present

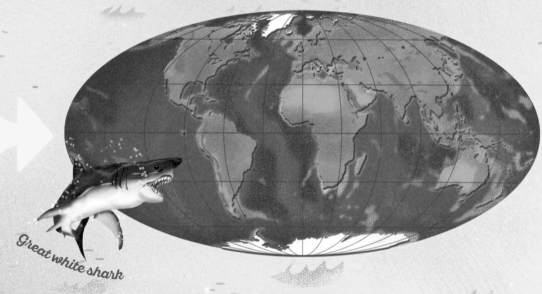

Great white shark

With the disappearance of the dinosaurs, the Earth as we know it gradually appeared, with its four large continental masses separated by three principal oceans dotted with islands and interior seas. The Arctic Ocean crowned it all, enclosed by the northern boundaries of the continents.

The next 250 million years

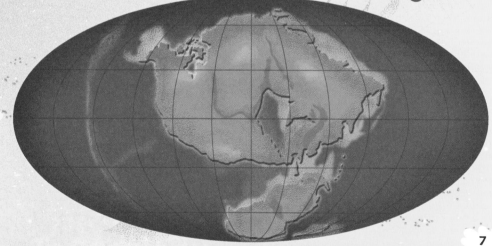

Our planet is always changing. Some scientists believe that over time, Africa will collide with Europe, the Mediterranean Sea will become closed off, and America will touch Asia. And within 250 million years, perhaps, only two continents will remain, with a single, immense Pacific Ocean.

ARCTIC

NORTH AMERICA

Northern Atlantic Ocean

Northern Pacific Ocean

CENTRAL AMERICA

Map of the World

SOUTH AMERICA

Southern Pacific Ocean

Southern Atlantic Ocean

Arctic Circle

Arctic Ocean

✛

NORTH POLE

Antarctic

Arctic Ocean

EUROPE

ASIA

Black Sea

Mediterranean Sea

Northern
Pacific
Ocean

AFRICA

OCEANIA

Equator

Indian
Ocean

Antarctic

AUSTRALIA

Antarctic Circle

SOUTH
⊕ POLE

Southern Pacific Ocean

Ocean

Ocean

ANTARCTICA

Ocean Vocabulary

Abyss

A very deep sea. Any area between 6,560 and 19,685 feet (2,000 and 6,000 m) deep is defined as "abyssal."

Abyssal fan

Fan-shaped underwater mass, formed by sediments (small pieces of material usually made of rocks and sand), that large rivers transport into the sea.

Abyssal plane

The flat area of the ocean floor that extends at depths from 13,123 to 19,685 feet (4,000 to 6,000 m).

Archipelago

A group of islands classified under a single name, such as the Aeolian Archipelago or the Malay Archipelago.

Basin

A huge hollow or dip in the Earth's surface filled by an ocean. Ocean basins cover about 70 percent of the Earth's surface.

Bay

A shallow stretch of ocean that opens up along the coast and is surrounded by solid land.

Benthic

Everything related to the ocean floor and the organisms that inhabit it.

Blue thermometer

This color generally indicates temperatures below or close to 32°F (0°C).

Compass rose

An illustration that shows the direction (North, South, East, West) on a map.

Current

The movement of seawater.

Delta

A wetland area that forms when a river empties out into a larger body of water. Deltas usually have a fan shape.

Depression or Trench

The deepest region of the ocean, such as the Mariana Trench.

Escarpment

A steep slope on the ocean floor.

Estuary

Where a river meets the sea or ocean.

Habitat

The specific environment in which an animal or plant species can live.

Ice shelf

A platform of floating ice that forms in the winter at the North and South poles, when the sea freezes over.

Mid-ocean ridge

The rising up of the ocean floor that separates two underwater basins.

Pelagic

Everything related to the open ocean.

Plateau

The flat underwater region that rises up from the surrounding ocean floor.

Planktonic

Everything related to plankton. Planktonic organisms like jellyfish are capable of small movements and shift with the currents.

Red thermometer

This color generally indicates temperatures greater than 32°F (0°C), typical of temperate and tropical oceans.

Ridge

An underwater mountain chain that rises up in the middle of the ocean.

Sea

A body of salt water smaller than an ocean that is surrounded by landmasses.

Strait

A narrow channel of water between landmasses that connects two seas.

Underwater canyon

A narrow, deep valley in the ocean floor.

EUROPE

Atlantic Ocean

Danube River

Northwest Wind

North Wind

Po River

Adriatic Sea

Rhône River

Ligurian Sea

Tyrrhenian Sea

Westerly Wind

Cold Current

Erbo River

Warm Current

Cold Current

Atlantic

Alboran Sea

Cold Current

Europe

Asia

MEDITERRANEAN SEA

Africa

AFRICA

Southwesterly Wind

Mediterranean Sea

Link to download Mediterranean Sea maps:
www.blackdogandleventhal.com/MapsoftheWorldsOceans/Mediterranean_Sea

Mediterranean Sea
Culture and Biodiversity

The Mediterranean Sea is located between three continents: Africa, Europe, and Asia. It has been a natural point of encounter for the populations inhabiting its shores for hundreds of thousands of years.

Three hundred thousand years ago, small groups of human beings lived in caves all around the Mediterranean Sea, from Spain in the west to Lebanon in the east. Stone objects from 130,000 years ago were found on the island of Crete, showing that people already knew how to navigate, or sail across the sea.

Five thousand years ago, some of the world's first civilizations and cultures, like the Egyptians, Cretans, and Phoenicians, began forming along the Mediterranean Sea's shores. Since they already knew how to navigate, these different groups of people were often in contact with each other. They exchanged goods, talked about ideas, and sometimes fought with each other.

The early civilizations of Greece created ports that still exist, such as Piraeus, the port of Athens, and the large port of Barcelona. The Romans relied on the Mediterranean so much that they called it *mare nostrum*, "our sea."

The name *Mediterranean* actually means "sea between the lands." It was first named that in the early Middle Ages, around 1,500 years ago, when a new culture, Islam, appeared from the Arabian Peninsula in the Middle East. Over time, Islam also left its mark on the Mediterranean region, from Morocco to Turkey.

There are many shipwrecks around the Mediterranean Sea that are still full of goods, some of which are thousands of years old.

Every region of the Mediterranean has different kinds of ships, such as the Gaeta in the Adriatic, or the Taka in the Black Sea. Some of them are beautiful ships made out of products found in the sea, like mother-of-pearl or coral.

Today, there are vessels like oil rigs and cargo, military, and research ships that sail all over the Mediterranean Sea. There are also many ferryboats and cruise ships—some as large as city blocks!

Today, more than ever, the Mediterranean Sea connects people and cultures from all over the world.

Six million years ago, the Mediterranean Sea did not exist. The Strait of Gibraltar was closed off and almost all the water of the Mediterranean had evaporated. The reopening of Gibraltar allowed the Atlantic Ocean to fill up the Mediterranean that we know today.

The Italian peninsula divides the Mediterranean into two parts: the colder western Mediterranean, and the warmer eastern Mediterranean. Species like the rabbitfish, originally from the Red Sea, live in this region.

The Mediterranean flows through the Dardanelles in Turkey and forms the Black Sea. The Black Sea's waters also come from large rivers like the Danube, making it less salty in areas than the Mediterranean. The less salty waters mean that fish like the European sea sturgeon can live there.

While it is small compared to oceans, the Mediterranean Sea is extremely rich in fish like sea bass and sea bream, which are eaten all over the world.

Posidonia, sometimes called sea grass, is an important type of sea plant that is found in the Mediterranean. It protects the coast from erosion and helps with the reproduction of fish, mollusks, and crustaceans.

Because the Mediterranean is so important to the people around it, and the entire world, there are many areas that are protected. For example, the Pelagos Sanctuary for Mediterranean Marine Mammals covers more than 32,000 square miles (84,000 km^2) and it helps protect rorqual whales, sperm whales, and dolphins.

There are many long, beautiful beaches on the Mediterranean coast, as well as vast, rocky stretches with high cliffs that provide protection to many seabirds, such as the rare Eleonora's falcon.

The coasts of the Mediterranean Sea attract millions of tourists each year to visit its beautiful beaches and explore its rich history. The area is also full of rare and protected species, like the loggerhead sea turtle and the monk seal.

Many scientists have studied the Mediterranean Sea over the years. In 1872, the Stazione Zoologica, one of the most important research institutes in the world, was created in Naples, Italy. They tested the Bathyscaphe *Trieste* in waters off the island of Capri, on August 26, 1953, which later made it possible to explore the extremely deep Mariana Trench.

Western Mediterranean Sea

SPAIN

MADRID ☆

Europe

BLACK SEA

MEDITERRANEAN SEA

Asia

Africa

Ebro Delta

Valencia

Júcar River

Valencia Deprecion

Balearic Sea

Ibi

Formen

Guadalquivir River

Segura River

Murcia

Cartagena

Cape Palos

Malaga

63°F (17°C)

Almeria

Capo di Gata

Alboran Sea

Eastern Alboran Basin

Gibraltar

Strait of Gibraltar

Ceuta

Tangiers

Western Alboran Basin

Cape Three Forks

Orano

MOROCCO

ALGERIA

FRANCE

Montpellier

Rhône
Delta

Rhône River

MONTE CARLO

Nizza

Marsiglia

Saint-Tropez

Ligurian
Sea

Gulf of Lion

Porquerolles
Island

Cap de Creus

Le Levant
Island

Ligurian-Corsican Basin

ANDORRA

Corsica

AJACCIO

Barcelona

Bonifacio

Strait of Bonifacio

Asinara
National Park

Gulf of Asinara

Balearic Bay

Porto Torres

Menorca

Alghero

Mallorca

Sardinia

70°F (21°C)

Plateau
Kene

San Peter
Island

CAGLIARI

Balearic Island

Mediterranean Sea

Sant'Antioco

Bay of Algiers

Sardino-Balearic
Abyssal Plain

Cap de Fer

ALGIERS

A L G E R I A

TUNISIA

17

Western Mediterranean Sea

 Flamingos have pink feathers because they eat brine shrimp. The birds are born with gray feathers that turn pink thanks to pigments in the tiny shrimp.

 In 2000, the rare **wreckage of an Etruscan ship** from 2,500 years ago was discovered near the Hyères islands in France.

 Deposits transported by the Ebro River have created an extensive area on the ocean floor, 30 miles (50 km) wide and more than 125 miles (200 m) long.

 Rhône river deposits, transported over millions of years, flow into underwater canyons carved out by the river and deposited over an area more than 6,560 feet (2,000 m) deep.

 The **stork** can build a nest 5 feet (1.5 m) wide and has a wingspan of more than 6 feet (2 m).

 Parasailing was created in the 1960s to train parachutists to launch without using an airplane. Today it is a popular tourist attraction on the Mediterranean Sea.

 The **common dolphin** can swim at speeds of up to 25 miles per hour (40 kph).

 The **Medes Islands Marine Reserve** was established in 1983 to preserve a part of the Mediterranean, where more than 1,200 species of animals and sea plants live.

 Jet-Skiing is one of the popular sports enjoyed by tourists visiting the Mediterranean.

 Weather buoys linked to the global GPS network provide information on air pressure, temperature, and the strength and direction of the winds.

 Red coral lives in the Medes Islands Marine Reserve off the coast of Spain.

 Hundreds of shipping containers travel on board large **cargo ships** that crowd the Mediterranean region east of the Strait of Gibraltar.

 The **Mediterranean gull** can move in flight, as well as on the ground, in search of food. It has a gray or black head and reddish feet.

 Windsurfing is another popular sports activity on the Mediterranean Sea.

 The **Egyptian vulture** migrates from Europe to Africa every year. The white bird can have a wingspan of nearly 6 feet (1.8 m).

 The **dusky grouper** is one of the most common fish in the Mediterranean. It can weigh up to 110 pounds (50 kg) and can live up to several decades.

 The **sardine** is sometimes called a blue fish since its upper body has a bluish color. It can live up to 15 years.

 Valencia, Spain, was home to a **storage facility for natural gas** that is now closed because it may have caused 500 earthquakes.

 Large sailing ships race in the **Mediterranean Tall Ships Regatta,** which runs from Barcelona (Spain) to Toulon (France) to La Spezia (Italy) every year.

 Paddling by **kayak** is an excellent way to admire the sea's beauty and life, especially in the warm waters between the islands of Corsica (part of France) and Sardinia (part of Italy).

 Anchovies are small silvery fish that live is schools (or groups) made up of tens of thousands. They swim between the surface of the water down to 490 feet (150 m) underwater.

The **spagnoletta** of western Sardinia is a fast and trustworthy boat that is used for fishing and other sea activities.

 In 1966, an American bomber with four **hydrogen bombs** on board collided with another airplane above the Spanish coast. Three were found, but one bomb was not recovered from the sea until two and a half months later.

 The **Barbary pirates** looted ships in the Mediterranean up until 200 years ago. Their name comes from the Berbers, an indigenous group of people from North Africa.

 The waters of western Sardinia are great for **scuba diving** thanks to the red coral that live in sheltered areas.

 Scientists use **ROVs**, or remotely operated vehicles, that can function underwater using a video camera and mechanical arms to study the Mediterranean Sea.

 Mother-of-pearl is a substance found in the shells of mollusks like clams and mussels. For centuries, mother-of-pearl from the African coast was used to decorate magnificent objects made of wood.

 The **submerged Almeria Canyon** was carved out by rivers about 6 million years ago, when the Mediterranean was almost dry. It is more than 30 miles (50 km) long and 3,900 feet (1,200 m) deep.

 One way to safely experience a thrilling, fast ride is the **banana boat**, an inflatable raft used in tourist locations around the Mediterranean, like Tunisia.

 Ancient ships like **triremes** were designed to navigate along the coast and not on the open seas.

 According to Greek mythology, the beautiful nymph **Calypso** imprisoned the hero Odysseus with a love spell for seven years on the island of Ogygia. Some believe this mythical island was on the Spanish coast.

 Two thousand years ago, courageous Phoenician navigators crossed the **Pillars of Hercules**, which is the narrow piece of land that connects the Atlantic Ocean to the Mediterranean Sea.

 Some wealthy Carthaginians, descendants of Phoenician colonizers, were buried in stone **sarcophaguses** sent by sea from Phoenicia or Egypt.

 Cruise ships as long as three football fields and higher than 20-story buildings travel routes in the Mediterranean.

 The **orca**, also known as a *killer whale*, is related to dolphins. It can weigh up to 12,000 pounds (5,443 kg).

 Mediterranean sardines from the African coast are often canned with **harissa**, an extremely spicy sauce made with a chili pepper typical of northwest Africa.

 The **Mobula**, or devil fish, is a large ocean-dwelling fish similar to the manta ray. It has a disk-shaped body that can be up to 17 feet (5 m) wide and lives in deep waters.

 The **monk seal** is one of the rarest sea mammals in the world and the only seal in the Mediterranean. The largest colony of Mediterranean monk seals reportedly lives along the coasts of Morocco.

 The **silky shark** enters the Alboran Sea through the Strait of Gibraltar, searching for little tuna, mackerel, and squid to eat.

 Doñana National Park, one of the largest and most interesting nature reserves in Spain, protects a great variety of environments and animal species, including thousands of water birds.

 The bottom of the **Alboran Sea**, east of the Strait of Gibraltar, is located at the center of a submerged mountain chain, the top of which is visible as Alboran Island.

 The **Doñana fossil dunes** were formed by sand collected and flattened by the wind. The dunes can be more than 65 feet (20 m) high.

 Laminaria, or kelp, is a type of large, dark brown algae that lives in the Alboran Sea and the Mediterranean Sea.

Ionic Sea

Ionic Basin

Strait of Messina

Reggi...

Sicily

Messina

Palermo

Trapani

Catania

Etna Volcano

Siracusa

Strait of Sicily

Pantelleria

Cape Bon

Cape of Tunis

Hammamet

Gulf of Hammamet

TUNIS ★

TUNISIA

Sfax

Gulf of Gabes

Galite Islands

Malta Plateau

Maltese Islands

Gozo Malta

Linosa

Pelagie Islands

Lampedusa

Tunisian Plateau

Kerkennah Islands

Djerba Island

Mediterranean Sea

Mediterranean Sea

63°F (17°C)

Sirte Abyssal Plain

Gulf of Sirte

Benghazi

Sirte

Misrata

TRIPOLI

LIBYA

AFRICA

Central Mediterranean Sea

AFRICA

21

Central Mediterranean Sea

 The **Venetian Lagoon** is the largest wetland in the Mediterranean.

 The **Po Delta** forms a series of intricate canals that connect the waterway with the open sea. Various species of plants and animals live here.

 The navigator **Christopher Columbus** set sail from Spain for the East Indies (South Asia) on August 3, 1492. He landed on San Salvador in the Bahamas on October 12, 1492 instead.

 Caulerpa taxifolia, a type of seaweed new to the Mediterranean, is known as "killer" algae because it has spread rapidly along the coasts and prevents native plants from growing.

 The **Aquarium of Genoa** was built in 1992 and is both the largest aquarium in Italy and one of the most comprehensive in the world. It houses approximately 12,000 examples of 600 different species.

 The European eel is born deep in the sea and then moves to the European coast when it is young. It then migrates all the way back to the Sargasso Sea to reproduce.

 The **pilot whale** is a large, dark-colored cetacean that belongs to the dolphin family. It is a skilled hunter of squid and can eat up to 220 pounds (100 kg) of them in a day.

 This mosaic from 1,500 years ago shows the ancient **port of Classe**, which was founded between 35 and 12 BCE by future Roman emperor Augustus.

 The **thresher shark** has a long tail that takes up almost half its body length. It allows this mammal to leap high above the water.

 The **Pelagos Sanctuary for Mediterranean Marine Mammals** protects cetaceans (sea mammals like whales and dolphins) between Italy, Monaco, and France.

 The **dusky grouper** is unusual because it changes gender as it grows. It is born female but when it turns about 12 years old, it changes into a male.

 Elegant lateen sails distinguish the **Gaeta**, a beautiful fishing vessel typical of the Adriatic, designed in the Italian city of Gaeta, on the Tyrrhenian Sea.

 At night in the Adriatic Sea, fishing boats are equipped with **lampara nets** that have powerful lights to attract plankton to the surface. The plankton then draw schools of sardines for the fishermen to catch.

 The **giant red shrimp** is a much sought-after crustacean, easily identified by its scarlet-red coloring. It lives at depths between 650 and 2,625 feet (200 and 800 m).

 The **bottlenose dolphin** can reach 13 feet (4 m) long and weigh up to 770 pounds (350 kg). It lives near the coasts and can often be seen leaping and flipping above the water.

 The **fin whale** is the largest cetacean in the world after the blue whale. It can be up to 65 feet (20 m) long and weigh 77 tons (70 metric tons).

 The **elephant shark** has a snout that looks like an elephant's trunk when it is young. It can reach up to 4 feet (1.2 m) long and feeds only on plankton.

 The **Bathyscaphe Trieste** is a small vehicle that can operate deep underwater. It was originally designed by Auguste Piccard and first immersed offshore of the island of Ponza, Italy, where it descended 10,335 feet (3,150 m).

 The **Aquabike World Championship race** often has a stop in the waters of Puglia, where Jet Skis can reach speeds of up to 105 miles per hour (170 kph).

 The **anchovy** has a tapered and cylindrical body, greenish-blue on the back and silvery-white on the belly.

 The **swordfish** has a long, pointed beak that it uses to defend itself and to spear the small fish it wants to eat.

 Because of its branchlike shape, the **red coral** was long considered a plant until biologists discovered that it was actually an animal with a limestone skeleton.

 When a **lobster** feels that it's in danger, its solid, fan-shaped tail folds back, allowing it to flee quickly.

 The **Eleonora's falcon** was named after Eleanor of Arborea, a Sardinian queen.

 The **sperm whale** can dive 7,380 feet (2,250 m) below the sea's surface and stay there for up to two hours.

 Sardines are named after the Island of Sardinia. They are small fish, 6 to 12 inches (15 to 30 cm) long, that feed on small crustaceans, mollusks, and eggs of other fish.

 Diving champions have set records off the island of Ischia, reaching depths of 650 feet (200 m).

Aquaculture, or fish farming, is the raising and fishing of sea organisms, such as fish and shellfish, on special farms.

 Flamingos can be up to 5 feet (1.5 m) tall and have a wingspan up to 5 feet (1.5 m). They live in the desert and very salty zones.

 In 1933, eight squadrons of Savoia-Marchetti S.55 seaplanes took off from the Mediterranean to make a famous **transatlantic flight** to Chicago and back.

 The **ocean sunfish**, also known as a **Mola**, is a strange-looking fish. It's shaped like a disk and doesn't have a tail. Ocean sunfish like sunning on the surface of the ocean.

 The **Marsili** is a still-active undersea volcano that is more than 9,800 feet (2,990 m) tall.

 Offshore wind turbines, which have been used to produce energy in northern Europe for more than 20 years, have also begun appearing in the Mediterranean. Some may reach to 325 feet (100 m) or taller.

 The **white coral** of the Mediterranean thrive in cold waters, where temperatures barely rise above 50 degrees (10 degrees C).

 Posidonia oceanica, or Neptune grass, is a sea plant, but it has roots, a trunk, leaves, flowers, and fruit, much like a plant you'd find on land.

 The Sicilian city of Trapani is famous for the **hand-carving of coral**, which has been practiced since the fifteenth century, using red coral pulled from the waters of the Mediterranean.

 Laminaria rodriguezii, or kelp, is a brown algae that grows deep in the sea along the Strait of Messina between the mainland of Italy and Sicily.

 The cliffs along the Strait of Messina was home to **Scylla**, a beautiful nymph of Greek mythology who was transformed into a six-headed monster by the sorceress Circe, who was jealous of her.

 The **loggerhead sea turtle** is a protected species that nests on Rabbit Beach on the island of Lampedusa, which is part of Italy.

 Sea sponges have been fished from the Mediterranean Sea since ancient Greek and Roman times.

 Scuba divers in the waters around Sicily can explore underwater caves and the ruins of ancient ships.

 Scientists believe **great white sharks** in the Mediterranean arrived from Australia tens of thousands of years ago.

 The **Stenella**, or striped dolphin, is about 8 feet (2.5 m) long and has a slender body. It lives in groups and often swims very close to ships.

 The **Balearic shearwater** is a seabird that nests on the islands of the Mediterranean and goes back and forth between the Mediterranean and the Black Sea.

 A stretch of the **Mattei** (or Trans-Mediterranean) **Pipeline**, which brings natural gas from Algeria, runs along the ocean floor between Tunisia and Italy for 1,370 miles (2,200 km), 235 of which are underwater.

 The **tuna**, which can be more than 10 feet (3 m) long, is one of the most sought-after fish in the Mediterranean.

 The **Scopoli's shearwater** is a seabird that moves throughout the Mediterranean and reproduces inside cliff grottos. At night, its cries sound like those of a newborn baby.

 The **Christ of the Sailors** is a statue 115 feet (35 m) under the sea that was placed there to commemorate Pope John Paul II's 1990 visit to Malta.

 The **beaked whale** is one of the most mysterious cetaceans in the Mediterranean. Although it can be up to 43 feet (13 m) long, it is difficult to find because it lives in deep waters far from the coasts.

 Significant oil deposits, found **along the coast of Tunisia**, are beginning to be fully used, like those of neighboring Libya.

 The **Mobula**, or devil fish, is similar to the tropical manta ray. It lives in the open sea, where it moves about in search of swarms of plankton, particularly tiny shrimp.

Many different kinds of vessels like 1,150-foot-long (350 m) **cargo ships** cut across the Gulf of Naples.

 The **bluntnose sixgill** is a large shark, about 25 feet (8 m) long, that lives in deep waters because it is very sensitive to light.

ITALY

Southern
Adriatic Basin

Brindisi ◇

◇ Taranto

Gulf of Taranto

Strait of
Otranto

Salonicco

Taso

Samothrace
Plateau

Samothrace

Gulf of
Salonika

Thracian Sea

Lemno

Aegean Sea

Corfù

GREECE

Crotone ◇

Catanzaro ◇

Leucade

Lesbo
Chio

Euboea

Cefalonia

Ionic Islands

Zacinto

Patras ◇

Corinto ◇

ATHENS

Andros

Ionic Sea

Peloponnesus

Plateau
Cyclades

Cyclades

Santorini

*Ionic
Basin*

East Mediterranean Ridge

Ptolemy Trench

Kythira

Kythira

Crete Sea

Hellenic

Trench

Crete

Gaudo

Sirte Abyssal Plain

Benghazi ◇

L I B Y A

Gulf of Bomba

Gulf of
Sollum

Gulf of Sirte

24

Eastern Mediterranean Sea

Sea of Marmara

Istanbul

Dardanelles Strait

ANKARA ✪

TURKEY

Europe

BLACK SEA

MEDITERRANEAN SEA

Asia

Africa

Izmir

Samo

Eğirdir Lake

Beyşehir Lake

Bodrum

Southern Sporrades Islands

Rhodes

Dodecaneso Islands

Karpathos

Antalya

Antalya Basin

Adana

Mersin

SYRIA

Ortones River

Cape Arnaoutis

Nicosia ✪

Cyprus

Cyprus Basin

LEBANON

BEIRUT ★

82°F (28°C)

Herodotus Basin

Mediterranean Sea

ISRAEL

JORDAN

TEL AVIV

Jerusalem

Dead Sea

Alexandria

Nile Delta

Port Said

Suez Canal

E G Y P T

Eastern Mediterranean Sea

 Large *aircraft carriers* cut across the Mediterranean, where important naval military installations like the ports of Naples and Taranto are located.

 Majestic temples and palaces are a sign of *Atlantis*, the mythical island that, according to Plato, was punished by Poseidon and swallowed up by the sea in a single day and night.

 Until the last century, a traditional activity in the Aegean Sea was the collection of sponges, carried out by *deep-sea divers*.

 The Dardanelles is a piece of land between Asia Minor and Europe. It was a thriving *trade route* for thousands of years.

 The giant *red shrimp* lives on ocean floors of compact mud known as *shrimp bottoms*.

 The *butterfly ray* has an unusual shape since its width is greater than its length. When it swims, its fins beat like the wings of a butterfly.

 Chimaera are also known as ghost sharks since they live in deep waters and are difficult to see.

 Enormous *ferries* departing from the famous port of Piraeus in Greece transport vehicles and goods through the Aegean Sea, where each island must receive supplies from the mainland.

 Statuettes with long-horned helmets represent the *Sea Peoples*, who invaded the Mediterranean from the east about 3,200 years ago.

 According to Greek mythology, *Poseidon* is the god of the seas, although he was "born" as lord of the soil, earthquakes, storms, and horses.

 Cruise ships on the Mediterranean can carry up to 6,000 people—which is the size of a small city.

 According to Greek mythology, *Sirens*, who enchanted sailors with their song, were among the creatures of the ancient Mediterranean. They may have been dolphins that communicate using a "musical" language.

 The *monk seal* gets its name from the brown color of its coat, which resembles the clothing worn by monks. They are slow on land, but move extremely fast in water.

 Minoan craftsmen often depicted octopuses with long winding tentacles on vases created 4,000 years ago.

 The *nomad jellyfish* is a gigantic jellyfish that originated in the Indian and Pacific oceans. It has a powerful sting, and is now spreading through the Mediterranean.

 Oil platforms are a common sight in the Mediterranean.

 The *grampus*, or Risso's dolphin, is small-to-average-sized and can grow up to 11.5 feet (3.5 m) long and weigh 880 pounds (400 kg). Unlike other dolphins, they have no upper teeth.

 Scuba diving is popular along the Turkish coast, where divers can see rich marine life, along with ancient shipwrecks.

The transparent and warm waters of the Ionian Sea are popular with *scuba divers* searching for octopuses, barracudas, moray eels, tortoises, parrotfish, and more.

 Some have suggested that the mythical island of Atlantis is actually the ancient island of *Thera* (now called Santorini), which mostly disappeared into the sea after a catastrophic eruption more than 3,500 years ago.

 The *Uluburun Shipwreck* was discovered in 1982 off the coast of Kaş, Turkey. It is believed to be nearly 3,500 years old!

 Pumice stone is usually found in volcanic regions. It can be seen floating in certain areas of the Aegean Sea.

 Intense *cargo ship* traffic cuts through the waters of Egypt toward coastal trading ports, such as Damietta, Port Said, and the Suez Canal.

 The *soft-shell turtle* has become extremely rare and is a protected species. It lives in many rivers that flow into the Mediterranean, from Egypt to Turkey.

 The *argonaut* is a mollusk that is related to the octopus and the squid. The male is very small, 0.4 inch (1 cm), while the female is larger, 8 inches (20 cm).

 According to Greek mythology, *Gorgons*, like Medusa, were monsters that lived on an island in the Mediterranean and could kill people by just looking at them.

 The ancient city of *Heracleion* in Egypt was submerged 1,500 years ago by either tidal waves or floods.

 More than 80,000 *fishing boats* are active in the Mediterranean Sea.

 The *scorpionfish* has large fins that look elegant and soft, but has spines underneath that can sting with dangerous venom.

 Cargo ships transport goods along the Sea of Marmara, an access route to the Mediterranean that is vitally important to the countries that border the Black Sea.

 Dolphins can be found all around Greece. In ancient Greek texts and myths, they often rescue shipwreck victims.

 The *loggerhead sea turtle* finds locations to reproduce on the large beaches along the coasts of the southern Mediterranean, which are less populated than other areas.

 Great white sharks are found in the warmer waters between Greece and Turkey.

 Rabbitfish and other tropical fish arrived in the Mediterranean from the Suez Canal in Egypt.

A Sea to Be Discovered

The Mediterranean Sea has been home to ancient civilizations like the Egyptians, Greeks, and Romans. People have been exploring the sea's underwater depths for thousands of years and have enjoyed swimming with the groupers, seahorses, and moray eels while being surrounded by fantastic landscapes.

Like other waters of the world, the Mediterranean Sea is always evolving. Today, climate change is bringing tropical species from other parts of the world that are altering the sea's underwater landscapes.

Black Sea

MOLDOVA

ROMANIA

BULGARIA

TURKEY

U K R

Odessa

Dnieper River

Karkinit Bay

Tarkhankut Peninsula

Crimea

Sevastopol

Chersonesus

Cape Aya

Danube Delta

Danube River

Constanța

Kaliakra Peninsula

Varna

Burgas

Emine Peninsula

Cape Iğneada

Viteaz Canyon

Bosphorus Canyon

Euxine Abyssal
7,400 feet

Cape Krempe

Cape Baba

Zonguldak

Cape Pazarbasi

Istanbul

Bosphorus

Sea of Marmara

Marmara

Lake İznik

The Dardanelles

T U R

Black Sea

 Large colonies of **pelicans** live in the Danube Delta and in the surrounding waters, where there are lots of fish that they swallow with their long beaks.

 The **beluga sturgeon** is the largest fish in the Black Sea and can live up to 100 years!

The first **steam-powered battleships** and warships were used on the Black Sea during the Crimean War (1853–1856).

 The **white-tailed eagle**, also known as the sea eagle, has a wingspan of about 8 feet (2.5 m) and catches fish close to the surface of the water by grasping them with its claws.

 The **sea walnut** is an almost transparent creature similar to a jellyfish. It lives in the Black Sea, feeding on fish eggs and small crustaceans.

Busts of various illustrious figures from the former Soviet Union were moved to the **Alley of Leaders**, an unusual underwater tunnel off the coast of Crimea, after the collapse of Communism.

 The **turkey vulture** has a brick-red head and black and white feathers. It nests in the arctic regions, but migrates to the Danube Delta to spend the winter.

 The **Dalmatian pelican** is one of the rarest birds in the Danube Delta and one of the largest pelicans on earth, with a wingspan up to 11 feet (3 m).

 The **bottlenose dolphin** has a rich vocabulary of whistles and sounds, which many believe could be an actual language.

 The Black Sea is popular with **sea kayakers**, who can easily explore the coasts and admire the beautiful landscapes.

 The **spiny dogfish** is a small shark, slightly more than 3 feet (1 m) long, which has sturdy thorns on its two dorsal fins.

 A **mermaid skeleton** was said to have been discovered off the Bulgarian coast, but it turned out to be a hoax.

 Oil deposits of the western Black Sea, tapped since the 1980s, are proving to be increasingly important and represent a growing resource for the region.

 Mullets live in coastal waters and estuaries near to the sea. They are sometimes called "happy" or "jumping" fish because they skip along the surface of the water.

 The **black whiting** has five fins and a small mouth. It feeds off plankton, fish, and crustaceans.

 The Black Sea does not have crystal-clear visibility or the colorful creatures of warmer waters, but **deep-sea divers** can make fascinating archaeological discoveries there.

 The **moon jellyfish** can grow up to 15 inches (40 cm) in diameter and have a pinkish, transparent umbrella that makes it possible to see its four-lobed internal organs.

 The **horse mackerel** is a deep-sea species that migrates northward in the Black Sea in the spring to reproduce and returns south in the autumn.

 The **Danube Delta** is where the Danube River meets the Black Sea. It covers an area of nearly 1,600 square miles (4,150 km²).

 Beluga whales are also known as white whales due to their color. They are actually gray when they are born and turn white as they get older.

 The Black Sea interests many scientists, and there are **research vessels** with cargo equipment, heliports, and laboratories all over its waters.

One area in the Black Sea is called the "Bermuda Triangle," because ships are swallowed up by vortexes and **whirlpools.**

 The **porpoise** is one of the smallest marine mammals in the world. It is around 4.5 feet (1.4 m) long and weighs only 88 pounds (40 kg). It lives in small schools in shallow coastal waters.

 The Black Sea has been broadly explored thanks to **ROUs** (remotely operated vehicles) that, in Romania, are also featured in an international competition dedicated to these remote-controlled vehicles.

 The **Black Sea Fleet**, which has patrolled the territorial waters of Russia for some 325 years, now employs modern vehicles, such as nuclear submarines.

 The dense waters of the Black Sea help to preserve ancient **shipwrecks**, which continue to be discovered on the ocean floor.

 The **turbot** is a round, flat fish that has hard bony knobs, or *tubercles*, on its back.

 Ultramodern warships, from *hovercraft missile launchers* to the *Kuznetsov* battle cruiser aircraft carrier, are built in Ukrainian shipyards.

 Bluefish are *pelagic* fish, which means that they live near the surface of the ocean. They are very good hunters, and pursue large schools of mullet, anchovies, and mackerel.

 The Greek mythology hero **Jason** crossed the Black Sea and confronted numerous dangers to capture the Golden Fleece, a mythical treasure that probably symbolized the region's wealth.

 The **Rapana**, or rapa whelk, is an exotic sea snail that arrived to the Black Sea from Japan. It hunts and eats other mollusks, like clams and oysters.

 According to some people, the tale of the **Great Flood** dates back 12,000 years, to the end of the last Ice Age, when the waters of the Mediterranean flowed back into the basin of the present-day Black Sea.

 The **Euxine abyssal plain**, from the ancient Greek name for the Black Sea, occupies the central portion of this sea and reaches a depth of 6,560 to 7,215 feet (2,000 to 2,200 m).

 Flyboarding is popular in the Black Sea, where enthusiasts can fly on a board propelled by powerful jets of water.

 Seagrass is sea plant that looks like grass. It grows on shallow and sandy ocean floors near to the coast.

 Thousands of tourists visit the Black Sea on **cruise ships** every year.

 The **taka** is a stable and dependable fishing and shipping vessel that can handle the dangerous waves of the Black Sea.

 The **anchovy** is the most commonly fished blue fish in the Black Sea. It feeds off small plankton organisms and occupies an important place in the food chain of this sea.

 The Black Sea provides ingredients for numerous recipes from the six different countries that surround it, such as Turkish **hamsi rice**, with anchovies and pine nuts.

 The **thornback ray** can grow up to 3 feet (1 m) long and has curved thorns on its back and on its stomach.

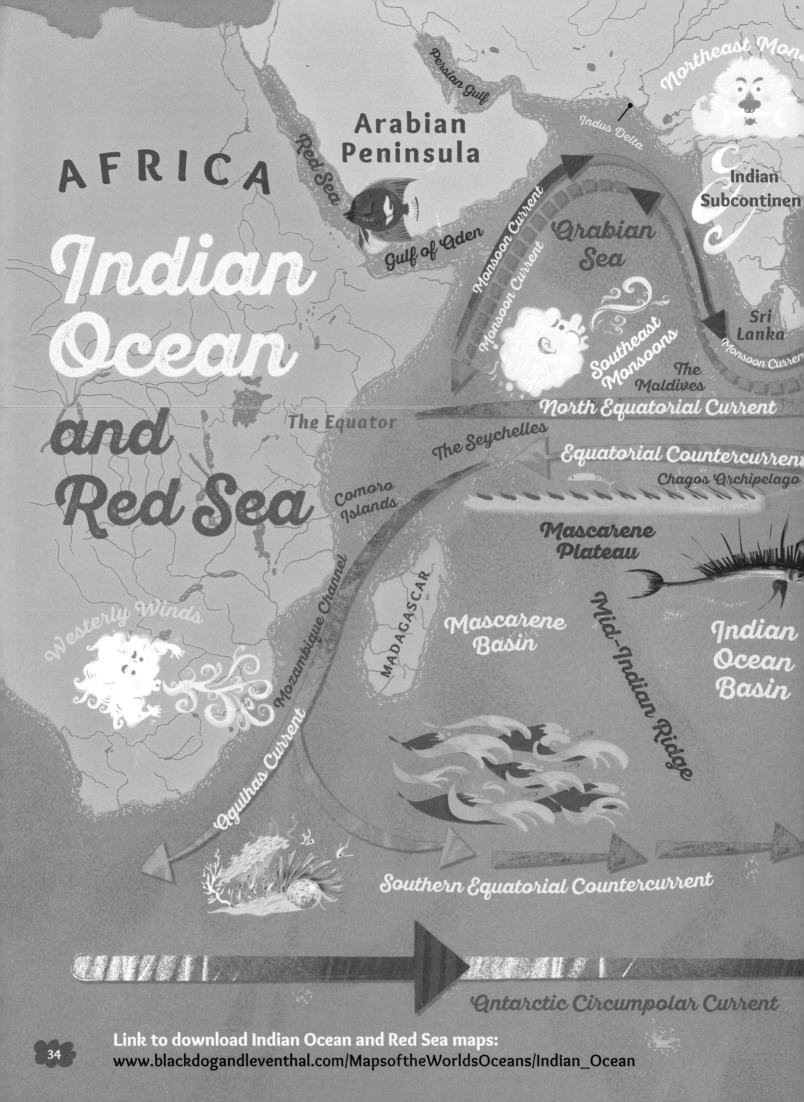

AFRICA

Indian Ocean and Red Sea

Persian Gulf

Arabian Peninsula

Red Sea

Indus Delta

Northeast Mon...

Indian Subcontinen...

Gulf of Aden

Monsoon Current

Monsoon Current

Arabian Sea

Sri Lanka

Monsoon Current

Southeast Monsoons

The Maldives

North Equatorial Current

The Equator

The Seychelles

Equatorial Countercurrent

Chagos Archipelago

Comoro Islands

Mascarene Plateau

Mozambique Channel

MADAGASCAR

Mascarene Basin

Mid-Indian Ridge

Indian Ocean Basin

Westerly Winds

Agulhas Current

Southern Equatorial Countercurrent

Antarctic Circumpolar Current

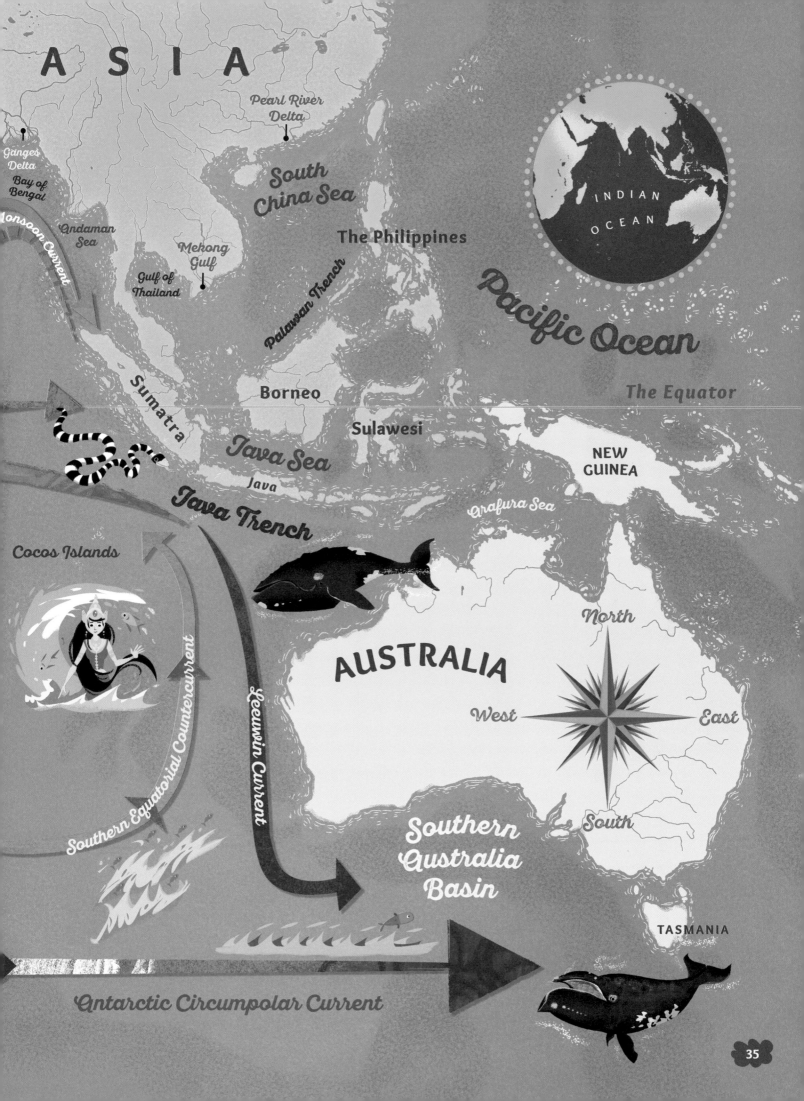

ASIA

Ganges
Delta

Bay of
Bengal

Monsoon Current

Andaman
Sea

Pearl River
Delta

South
China Sea

The Philippines

Gulf of
Thailand

Mekong
Gulf

Palawan Trench

Sumatra

Java Sea

Borneo

Sulawesi

Java

Java Trench

Cocos Islands

Southern Equatorial Countercurrent

Leeuwin Current

INDIAN
OCEAN

Pacific Ocean

The Equator

NEW
GUINEA

Arafura Sea

North

AUSTRALIA

West

East

South

Southern
Australia
Basin

TASMANIA

Antarctic Circumpolar Current

Indian Ocean
Merchants and Monsoons

About 100,000 years ago, some groups of *Homo sapiens* left Africa and traveled east to the Indian Ocean. It was the beginning of a 40,000-year journey that would take their descendants to Australia, at the opposite side of the ocean.

Those ancient explorers did not know how to navigate, but over time, their descendants learned to take advantage of the seasonal winds that blow from Asia to Africa and back again.

The Indian Ocean enters the Asian continent through the Red Sea, the Persian Gulf, and the Bay of Bengal. This has allowed for an easy exchange of goods and ideas and helped the Egyptians, the Sumerians, and the inhabitants of the Indus Valley create the earliest civilizations in history more than 5,000 years ago.

Trade of important resources like metals, gemstones, timber, and food helped shape the history of the Indian Ocean. It became known as a place of fabulous riches.

Because of its reputation for riches, it was the first ocean to be thoroughly explored by adventurous travelers, including the Greek Eudoxus of Cyzicus 2,100 years ago and the Arab Sulaiman Al Mahri in the sixteenth century, who spread the word in the West about routes to India.

Ancient kingdoms such as Srivijaya created a maritime empire in the eastern region 1,500 years ago, and were later replaced by empires like China, which established trade with Africa 600 years ago. Portugal, England, and Holland fought for control of the Indian Ocean for more than three centuries.

Today the Indian Ocean is still an extremely important commercial route for cargo ships and oil tankers between the East and West. Traditional fishing boats such as Arab and African dhows, catamarans from Southern India, and Indonesian outrigger canoes frequently ply the waters as well.

Fishing is still one of the Indian Ocean's most important sources of economic activity.
The ocean is also the site of research and exploitation of gas and oil deposits and the laying and maintenance of underwater cables that allow intercontinental communications.

After the terrible tsunami of 2004, the Indian Ocean has also been monitored by a seismic buoy system, set up to protect the local populations and the millions of tourists who visit fabled sites, including Mauritius, the Maldives, the Andaman Islands, and Indonesia.

The Indian Ocean is the third largest ocean in the world. It is the only one that does not connect to the Arctic seas, as it is closed off to the north by the gigantic continent of Asia. The large triangle of India lies at the ocean's center. Almost a continent, and its seas are inhabited by thousands of large and small species.

If you look at a map of the Indian Ocean, you will see that the Maldives and the Laccadive Islands are curiously aligned. India has shifted northward over thousands of years and some volcanic structures were left behind that created these two archipelagos.

The island of Madagascar in the Indian Ocean has only been inhabited by humans for a few thousand years. Eighty percent of its animal and plant species are indigenous to the island, which means they only live there. The rare coelacanth, for example, has been in the waters near Madagascar since dinosaurs were alive.

The northwestern portion of the Indian Ocean borders the desert that gave rise to the Red Sea, one of the most fascinating tropical seas in the world. It is rich in coral and fish of thousands of colors. The lands surrounding it are growing farther apart, and in the distant future, the Red Sea will become an ocean.

The Indian Ocean has always been a crossroads between East and West—and not only for people. Humpback whales, tuna, whale sharks, and great white sharks migrate in its waters from north to south and east to west.

Most oceans have large, stable wind currents. Since it is in a monsoon climate, the currents of the Indian Ocean change from season to season. In the winter, the currents run from India to Africa, and in the opposite direction in the summer. These changing winds impact the migrations of many fish, such as tuna.

Mangrove plants are shrubs or small trees that grow in muddy coastal areas. The coasts around the Indian Ocean are full of different species of mangroves thanks to the area's climate, currents, rivers, and rains. Mangroves play an important role in defending the coasts from erosion and protecting the coral reefs from coastal debris.

The Indian Ocean is known for its coral reefs, which are particularly abundant in the Maldives archipelago. These islands, which in some cases are less than 3 feet (1 m) above the sea, run the risk of being submerged as a result of rising seas caused by melting ice due to global warming.

Red Sea

SAUDI ARABIA

ISRAEL

JORDAN

Gulf of Aqaba

Sinai Peninsula

-332 m

Tiran Island

Sanafir Island

Sharm el-Sheikh

Gulf of Suez

Strait of Gubal

Hurghada

Ras Muhammad National Park

79°F (26°C)

Shaban Basin

Medina

Yanbu

Ras Banas

Quseir

Marsa Alam

Atlantis Trench

Eastern Desert

Nile River

Luxor

EGYPT

Aswan

Lake Nasser

38

Nubian Desert

Mecca

Jeddah

stop illegal fishing

Suakin Trench

Suakin Archipelago

Sudan Port

SUDAN

Nile River

YEMEN

SANA'A

Al-Hudaydah

Hanish Islands

Bab-el-Mandeb Strait

Gulf of Aden

DJIBOUTI

SOMALIA

Farasan Islands

86°F (30°C)

Dahlak Archipelago

Massawa

ERITREA

ASMARA

ETHIOPIA

Lake Tana

Nile River

Asia

INDIAN OCEAN

RED SEA

Africa

Red Sea

 About 50 ships can travel from the Mediterranean to the Red Sea through the *Suez Canal* every day.

 The *western reef heron* has dark gray feathers and lives in the rocky coasts of the Red Sea, where it feeds off the area's abundant sea creatures.

 The *blacktip reef shark* is one of the most common predators on the coral reef. It is easy to spot thanks to its black-tipped fins.

 The warm water in the Gulf of Aqaba, or Gulf of Eilat, at the northwest corner of the Red Sea allows *winter ocean swimming competitions* that are open to everyone, including professionals.

 There are many different kinds of *sea eagles* all over the world. They usually have disk-shaped bodies with long, narrow tails.

 The warm, calm, and transparent waters of the Red Sea are perfect for *aquatic sports* encouraged by the tourism industry, which is an important source of income for the surrounding countries.

 The *sohal surgeonfish* has a blue body with white stripes. It only lives in the shallow areas of coral reefs in the Red Sea.

 Long-beaked Stenella dolphins live in small schools and are very agile. They can leap up to 10 feet (3 m) above the water.

The pharaoh queen Hatshepsut sent long ships through the Red Sea to the *Land of Punt* 3,400 years ago to trade precious goods, such as ivory, with its people.

 In one of the most spectacular tales in the Bible, the Red Sea parts to allow the passage of the *Israelites freed by Moses*, during the reign of the pharaoh Ramesses II, approximately 3,300 years ago.

 The *Daedalus Reef* is one of the few coral *atolls*, or chain of islands, in the Red Sea. They were formed by a volcanic crater, and a small human-made island with a lighthouse lies at the center.

 The *Cassiopea*, or upside-down jellyfish, doesn't swim. It remains on the ocean floor, and its tentacles turn upward to absorb light and allow algae to grow in its body.

 Incredible *diving records* have been set in the waters of the Red Sea. Some dives have been more than 985 feet (300 m) below the surface and last many hours.

 The *Seascope* is a vessel with a transparent lower body that makes it possible to admire the wonders of the Red Sea with as little disturbance as possible to the environment.

 The *manta* is the largest ray in the world. Its body looks like it has two large wings, and it can be up to 23 feet (7 m) wide. It can weigh up to 3 tons (2.7 metric tons).

 Even though they are fitted with modern equipment, some *cargo ships* in the Red Sea still have the curved shape of traditional vessels.

 The *Spanish dancer* is one of the largest sea slug mollusks. They are usually red or orange with ruffled edges, and look like Spanish flamenco dancers when they swim.

 The Red Sea has some of the most famous *diving sites* in the world thanks to its abundance of sea creatures that create spectacular environments.

 The *guitarfish* has a flat shape and mostly lives on sandy ocean floors.

 Large *desalination plants* have been built along the arid Arabian coast of the Red Sea, making inland agriculture possible.

 The *hooded butterflyfish* has an orange head, a white body, and a black tail, and only lives in the Red Sea.

Non-governmental organizations fight *illegal fishing* in the Red Sea, which seriously harms the environment and the populations of the countries where this takes place.

Fire coral looks harmless, but its powerful sting can feel like burns on your skin.

The *bluespotted ribbontail ray* likes to spend time resting on sandy ocean floors. It looks harmless, with bright, neon-blue spots, but has poisonous spines on its tail.

Crown-of-thorns starfish are one of the largest sea stars in the world. They can grow to be 18 inches (45 cm) long and can have up to 21 arms.

The *royal angelfish* is bright and colorful, with vertical yellow and blue stripes.

The *Arabian angelfish* is a brightly colored blue and yellow tropical fish that lives in the Red Sea.

You don't have to go underwater to admire the wonders of the Red Sea. You can go *snorkeling*, swimming on the surface with a mask, fins, and snorkel.

The *whale shark* is the largest known cartilaginous fish, which means its skeleton is made of cartilage instead of bone. It can grow up to 40 feet (12 m) long and weigh up to 20 tons (18 metric tons).

The *oceanic whitetip shark* is a deep-water shark. It is one of the most dangerous of this species in the world.

Windsurfing is another classic sport practiced in the Red Sea, where the breeze blows in from the desert to the sea and back again, depending on changes in temperature.

The *roving coral grouper* is one of the most beautiful and colorful fish in the Red Sea. Its coat is bright red, studded with blue spots.

The *felucca* is a typical sailing ship of the Red Sea. It is a light and practical vessel, with a type of triangular sail that replaced the square sail used in ancient times, about 2,000 years ago.

The *Arabian angelfish* is a brightly colored blue and yellow tropical fish that lives in the Red Sea.

The *city of Suakin* in the Sudan was once an important center of trade between Africa and Arabia. Today it is surrounded by the continuous growth of coral on the surrounding ocean floor.

The *green sea turtle* eats algae and plants, which is why it is usually seen in underwater grasslands. It can rest or sleep underwater for up to five hours.

The salty water from the Red Sea is evaporated in *evaporation plants* to remove useful minerals.

In the 1950s, the oceanographer *Jacques-Yves Cousteau* began groundbreaking undersea explorations, showing the world the underwater wonders of the Red Sea.

Shipyards with dynamic positioning, which can maintain a fixed position during work, are used to put in structures and search for mineral deposits.

More than 2,000 years ago, the Greek navigator *Eudoxus* and his captain *Hippalus* arrived in India by sailing the southwest monsoon wind.

Oceanographic ships are floating laboratories that allow scientists to carry out long research "cruises" in the Mediterranean and Red seas.

The *giant moray* is an eel that can grow up to 10 feet (3 m) long. Giant morays are quick-moving and can chase prey, even among the most tangled coral.

The *Red Sea clownfish* relies on the stinging tentacles of the sea anemones it lives with to defend itself. In return, it defends the sea anemones from predators.

The *hammerhead shark*'s head, as its name suggests, is shaped like a hammer.

Red Sea: An Oasis of Coral

Incredible coral gardens live beneath the blue surface of the Red Sea, making it one of the most rich and colorful waters of planet Earth.

between Two Deserts

Arabian
Sea

Mascarene Plateau

79°F
(26°C)

Somali Basin

Socotra Island

The Seychelles

Gulf of Aden

Cape
Guardafui

YEMEN

Bab-el-Mandeb
Strait

SOMALIA

Horn of Africa

ERITREA

DJIBOUTI

MOGADISHU

Western
Indian
Ocean

KENYA

Malindi

Pemba Island

Zanzibar

MOMBASA

Victoria
Lake

DAR ES SALAAM

TANZANIA

Asia

INDIAN OCEAN

WEST
INDIAN
OCEAN

Africa

Indian Ocean

Mascarene Basin

Madagascar Basin

Madagascar Plateau

Tromelin

Mauritius

Réunion Island

Farquhar Atoll

Antsiranana

Antalaha

Antongil Bay

ANTANANARIVO

Manakara

Aldabra Island

Comoro Islands

MADAGASCAR

Toliara

Pemba

Nacala

Juan de Nova Island

Bassas da India

Europa Island

86°F (30°C)

Mozambique Channel

MOZAMBIQUE

Mozambique Basin

Malawi Lake

Zambezi River

Beira

45

Western Indian Ocean

 Captain Kidd was a sailor and pirate in the 1600s. Some people believe he buried a chest of gold and jewels on a desert island before he was caught and killed.

 Chinese admiral **Zheng He** made several expeditions to Africa about 600 years ago with fleets of sailboats called **junks**. Some accounts say the ships were up to 400 feet (120 m) long, twice as long as the largest European sailboats of that time.

 Boats with **lateen**, or **Latin rigs**, were popular during the Age of Discovery, from the fifteenth century to the eighteenth century.

 The **whale shark** is the largest fish in the world but it almost only eats tiny plankton.

 Pirogues are small boats from West Africa and Madagascar that are carved out of tree trunks.

 In 1498, the Portuguese explorer **Vasco da Gama** reached the coast of East Africa while traveling to India. He was the first European to arrive in India by passing through the Cape of Good Hope, the southern tip of Africa.

 The **sea coconut** is a symbol of the Seychelles. It is a coconut with two shells attached to each other, and contains the largest seed in the plant kingdom.

 The cuisine of the Seychelles islands is influenced by France, since it was once a French colony. One famous recipe is their **Creole grilled snapper.**

 Bull sharks are known as **Zambezi sharks** in Africa. They are considered the most dangerous sharks in the world.

 The waters of East Africa have strong currents and submerged rocks, making it difficult for ships to navigate. There have been many old **shipwrecks** found near the coasts.

 At low tide, Indian women and children collect **shells** from the shores of the Indian Ocean to sell and to use at home.

 Octopus fishermen on the island of Mauritius must have excellent eyesight, because their prey know how to camouflage themselves to perfection.

 Along the coast of the Horn of Africa, the eastern part of the continent that forms a peninsula, fishing is mostly still done by small communities of **fishermen.**

 Rubies and **walnuts** were considered precious goods that arrived to Europe via the Silk Road between Asia and Africa from the second century BCE until the fourteenth century CE.

 International artists have created magnificent **graffiti** on traditional boat sails on the island of Réunion.

Remora fish are also known as suckerfish because they attach themselves to ships and larger marine animals.

 Dhows are quick-moving and easy-to-handle sailing ships that have been used for centuries in the western Indian Ocean for fishing and shipping. Today they also compete in regattas.

 The **Réunion hotspot** continually emits lava that, over millions of years, has resulted in a great many islands located between the equator and India, such as the Maldives.

 The Indian Ocean is very important not only to those who live in the region but for trade all around the world. Patrol ships like those of the **Coast Guard** help to protect the waters.

 Every four years the islands in the Indian Ocean participate in the **Indian Ocean Island Games**, competing in sports like swimming, sailing, basketball, and judo.

 Seismographs are instruments used to measure earthquakes and have been installed underwater east of Madagascar to explain the "**Earth's Hum**," a mysterious murmur that may be produced by the movement of ocean waves.

 Some believe that Madagascar, Africa, India, and Australia were united by **Kumari Kandam**, a now-vanished continent shaped like an enormous triangle, thousands of years ago.

 Athletes have crossed the Indian Ocean from Australia to Mauritius in **rowboats**, traveling 3,100 to 3,700 miles (5,000 to nearly 6,000 km) in two or three months.

 The **Amur falcon** flies south from Siberia to Africa in the winter, passing over the Indian Ocean. It returns north in the summer.

 The Hammerhead shark uses its wide head to trap stingrays, which are its favorite prey.

 Japanese submarines, English ships, and French torpedo boats fought in 1942 in the little-known Battle of Madagascar during World War II.

 Lascars were sailors from the Indian Ocean area hired by European ships from the sixteenth century until the middle of the twentieth century to protect the waters.

 Sextants are navigational instruments used to measure the distance between two objects. They were developed about 290 years ago and were used to figure out the position of ships, which helped navigators explore the Indian Ocean.

 The **aspidochelone** is a fabled sea creature with trees and rocks on its back that was supposedly seen by sailors in the Middle Ages.

INDIA

Persian Gulf

Indus River Delta

Kathiawar Peninsula

Gulf of Kutch

Gulf of Khambhat

◇ Surat

◇ Mumbai

◇ Goa

◇ Mangalore

◇ Cochin

◇ Trivandrum

Cape Comorin

Chennai ◇

Coromandel Coast

Bay of Bengal

Palk Strait

Gulf of Mannar

SRI LANKA

COLOMBO ★

Laccadive Sea

Ceylon Abyssal Plain

Chagos-Laccadive Ridge

Laccadive Islands

The Maldives

MALÉ ★

72°F (22°C)

Gulf of Oman

MUSCAT ★

Ras al Hadd

Murray Ridge

OMAN

Masirah Island

Masirah Gulf

Arabian Sea

Arabian Basin

East Sheba Ridge

Socotra Island

Carlsberg Ridge

Somali Basin

The Equator

48

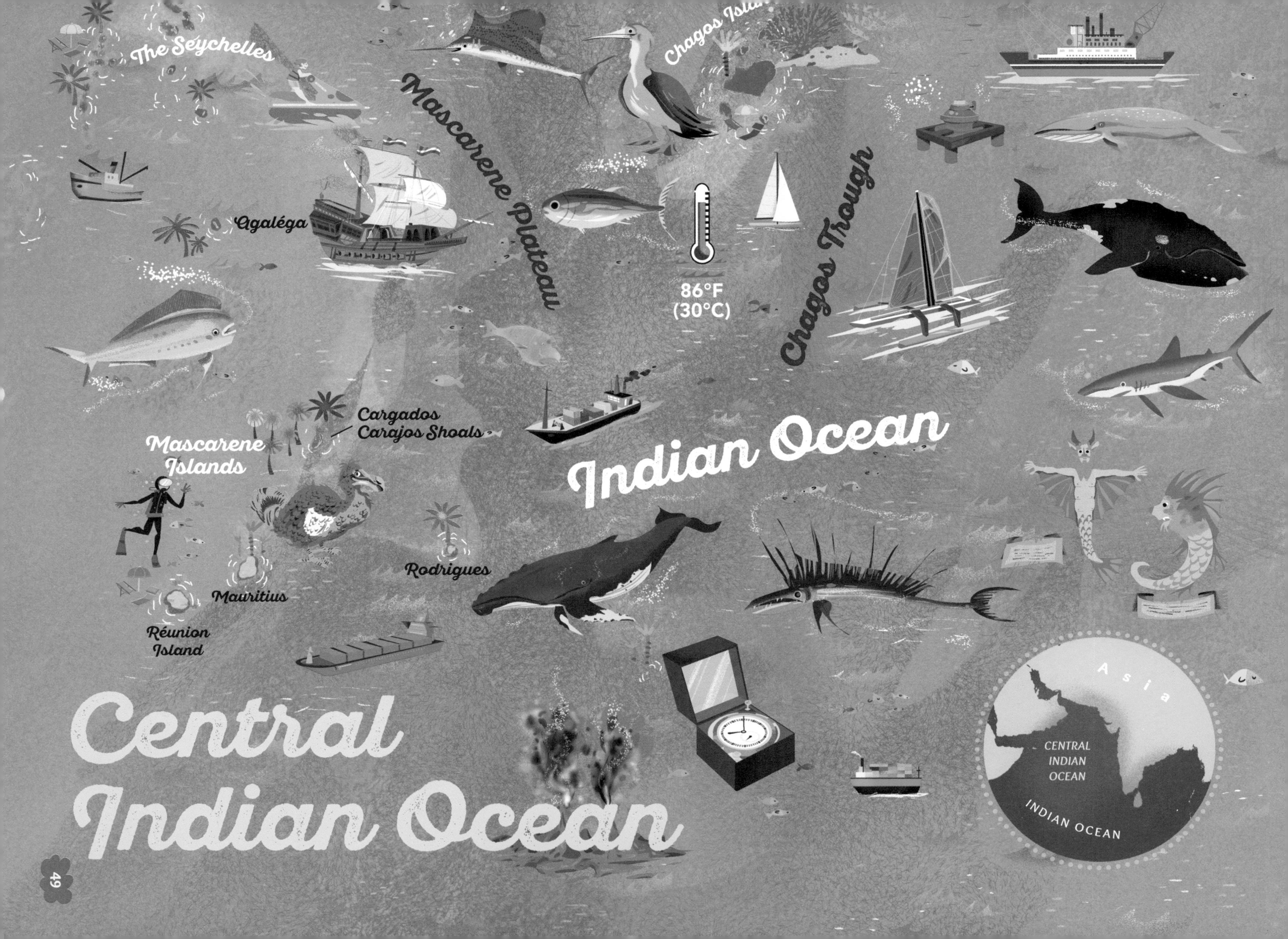
Central Indian Ocean
The Seychelles
Chagos Islands
Mascarene Plateau
Agaléga
86°F (30°C)
Chagos Trough
Cargados Carajos Shoals
Indian Ocean
Mascarene Islands
Rodrigues
Mauritius
Réunion Island
Asia
CENTRAL INDIAN OCEAN
INDIAN OCEAN

Central Indian Ocean

 The **Platanista**, or Indus river dolphin, is a freshwater dolphin that also lives in the Indus River Delta. It has poor eyesight, so it uses sonar to hunt in the murky waters.

 Moroccan scholar and explorer **Ibn Battuta** arrived in China around 700 years ago, traveling through the Indian Ocean a few years after Marco Polo's return to Venice.

 The **coconut crab** lives mainly on land and has extremely powerful claws it uses to cut holes in coconuts to eat the flesh inside.

 Ruins of a sunken city were discovered near the city of Dwarka, India. Some people believe it may be **Dvārakā**, the mythical residence of the god Krishna.

 Rays gather in the waters of Hanifaru, an uninhabited island of the Maldives, between May and November.

 For centuries divers fished **pearls** out of oyster shells in the ocean floors around Sri Lanka for jewelry like necklaces. Today, pearls are mostly cultivated in breeding grounds.

 A **kotiya** is a small wooden sailing boat that was used in Oman for centuries for sea trade between India and Arabia.

 The **southern right whale** only lives south of the equator. It grows up to 56 feet (17 m) long, can weigh up to 80 tons (73 metric tons), and has large white patches all over its body.

 Outrigger canoes are made from carved-out tree trunks attached to a stabilizer to prevent them from capsizing. They are still used in the Indian Ocean today.

 At night, tiny **bioluminescent algae** give off bright flashes of blue light when they are pushed by the waves onto the beaches, creating an extraordinary sight.

 Turbinella, or divine conch, are large seashells. They are sacred in the Hindu religion and are used to create elegant carvings and jewelry.

 An enormous fish is about to swallow up the boat of **Sinbad the Sailor**, a famous explorer of the Indian Ocean...although only in the tales of *One Thousand and One Nights*.

 Aquatic sports such as **Jet-Skiing** are one of the many attractions at tourist resorts in the Seychelles and the Maldives.

 In 2014, an **astrolabe** was found on board a ship from explorer Vasco da Gama's fleet that sank in 1503. An astrolabe is the oldest known navigational tool. It was used to determine the positions of the stars to help explorers and sailors figure out where they were.

 The **blackfinned anemonefish** lives among the tentacles of large anemones, which look like flowers but are actually animals.

 The island of **Panchaia** may have been somewhere between India and Arabia. According to legend, it was inhabited by a lost tribe of Greeks who built a large temple dedicated to Zeus on the island.

 A **kettuvallam** is a floating house made of bamboo wood and rope that was once used in southern India to transport rice and spices. Today it is used as a tourist vessel.

 Traditional Indian **catamarans** are boats made of simple wooden planks tied together. They are used by fishermen in southern India.

 Technological advances facilitate the discovery of new **hydrocarbon deposits** such as those offshore of Oman, dating back to the era of dinosaurs, 70 million years ago.

 The **grey heron** moves slowly in shallow water, and when it spots a fish, it quickly captures it with its long, pointed beak and a rapid twitch of its neck.

 The **SEA-ME-WE 4 cable** is the principal Internet connection between Asia, the Middle East, and Europe. It is about 11,680 miles (18,800 km) long and runs along the floor of the Indian Ocean.

The **whale shark** is a migratory shark that makes long stopovers in the waters of the Maldives to feed on the plankton that form there during monsoon season.

 The **yellowfin tuna** gets its name from the color of some of its fins.

 Humpback whales travel 4,350 miles (7,000 km) from the cold Antarctic waters to the Indian Ocean where they mate and reproduce.

 Stand-up paddleboarding is a perfect way to explore the transparent waters of the Indian Ocean.

 The **British East India Company** was created in 1600 and dominated sea trade with Asia until the late 1800s.

 Italian astrologer and cosmographer **Paolo dal Pozzo Toscanelli's 1474 world map** shows the Indian Ocean at the center, but North and South America are absent. They were discovered 18 years after his death.

 The **Omura's whale** was named for a Japanese scientist who was an expert on cetaceans. It is a small and very rare whale.

 The **humphead wrasse**, also known as the **Napoleon fish**, is a large coral reef fish that can grow up to 6.5 feet (2 m) long and has a big bulge on its forehead.

 In 1662, the Dutch ship **Arnhem** sank near a group of islands northeast of Mauritius. The survivors, who were rescued three months later, may have been the last people to see the now-extinct **dodo bird.**

 An **ancient sunken ship** from 2,000 years ago was found near the island of Sri Lanka. It is the oldest vessel ever found in the Indian Ocean and may have transported metal and colored glass to ancient Rome.

 Drilling ships are used for both scientific research and industrial purposes. Today, they can drill down more than 9,800 feet (3,000 m).

 The **marine chronometer** was invented by John Harrison in 1761. It was used on board ships to determine longitude, which helped the navigation of English sailboats in the Indian Ocean.

 The **Huvadhu Atoll** in the Maldives covers an area of 1,217 square miles (3,152 km^2). It has 255 islands, more than any other atoll in the world. It is home to large whale sharks, manta rays, and tortoises.

 The **red-footed booby**, named for the red color of its webbed feet, nests on the ground but hunts in the sea, flying as far as 93 miles (150 km) in search of fish.

 A **hydrothermal vent** is an opening in the surface of the ocean floor. Boiling water heated by volcanic gases escapes from them, heating the surrounding water and making it possible to catch sight of strange deep-sea creatures.

 Large openings in the coral reefs that surround the **atolls** of the Maldives allow the exchange of water inside the lagoons when the sea shifts.

 Underwater seismographs are in the Indian Ocean to help predict earthquakes and tsunamis.

 The **pygmy blue whale** is smaller than the blue whale, but still grows up to 80 feet (24 m) long and can weigh 90 tons (82 metric tons).

 Ultra-modern **trimarans** periodically race across the Indian Ocean, from Cape Agulhas in South Africa to Tasmania, with record times of less than one week.

ASIA

CHINA

Hainan

Gulf of Tonkin

Paracel Islands

South China Sea

Da Nang

VIETNAM

Nansha Qundao

Sunda Shelf

Borneo

INDONESIA

Java Sea

HANOI

Belitung

Bangka Island

CHINA

ASIA

INDIA

Calcutta

Ganges Delta

Chittagong

MYANMAR

LAOS

VIENTIANE

THAILAND

BANGKOK

Yangon

Irrawaddy Delta

Cheduba Island

Gulf of Martaban

Andaman Sea

CAMBODIA

PHNOM PENH

Ho Chi Minh

Gulf of Thailand

Ko Samui

Cà Mau Peninsula

Andaman Basin

MALAYSIA

MALAYSIA

KUALA LUMPUR

Malay Peninsula

Malacca Strait

SINGAPORE

Riau Archipelago

Lingga Island

Medan

Sumatra

Mentawai Islands

Sunda Trench

Palembang

Bengkulu

Bay of Bengal

Visakhapatnam

Andaman Islands

Nicobar Islands

Ceylon Abyssal Plain

Cocos Basin

Chennai

Palk Strait

SRI LANKA

COLOMBO

52

Semarang

Surabaya

Java Trench

Christmas Island

Gascoyne Abyssal Plain

AUSTRALIA

Shark Bay

Perth Basin

OIL

Investigator Ridge

Cocos Islands

Ninety East Ridge

Eastern Indian Ocean

Asia

SOUTH CHINA SEA

Australia

Australia

EASTERN INDIAN OCEAN

INDIAN OCEAN

Eastern Indian Ocean

 Cowry shells were once used as money in some countries around the Indian Ocean.

 In the Gulf of Siam, fishing boats often have long "wings" with powerful lamps, making it possible to carry out **nocturnal fishing**, or fishing at night, when the catch is particularly plentiful.

 The **harlequin sweetlips** grows up to 28 inches (70 cm) long and can weigh 15 pounds (7 kg). It has large lips that it uses to dig in the sand in search of prey.

 Marco Polo traveled to China overland but returned to Europe by sea and probably visited the island of Sumatra.

 Harpoon fishing is a traditional technique used on the Andaman Sea.

 The **giant barrel sponge** can be up to 6 feet (2 m) high and 3 feet (1 m) in diameter. Animals become encrusted on it and inside it, providing food and shelter for fish.

 The **Mergui Archipelago** is a natural paradise inhabited by the rare Asian small-clawed otter, large rays, sea eagles, and shark species such as the whale shark and the silvertip shark.

 Fish sauce is a specialty of the cultures around the Indian Ocean. It is made from salted fish and used instead of salt in countries like Thailand and Cambodia.

 In the Andaman Sea, **colorful scarves** hung from fishing and transport ships are good luck charms, but they also serve to identify their owner from a distance.

 The Moken are an indigenous group of people that live on the southeast coast of the Bay of Bengal and move from one island to another on board **sailing ships.**

 The Arab **Sulaiman Al Mahri** was one of the explorers who encouraged Islamic and European expansion in the region, making note of routes, ports, and other information during his voyages in the sixteenth century.

 A combination of Asian and European sailboats, the **gulet pinar** of Indonesia has been in use for 500 years. Its construction is so ingenious it is included on the UNESCO cultural heritage list.

 The **sawfish** has a pointed beak with sharp, pointed teeth along the sides. It uses sensors on its "saw" to locate prey.

 Chinese explorers like Zheng He went on many expeditions through the Indian Ocean to create new trading routes with other countries and cultures.

 The **snipefish** lives in deep waters between 3,280 and 13,120 feet (1,000 and 4,000 m) below the surface. Its slender body reaches 5 feet (1.5 m) long and its mouth looks like a long, thin beak.

 The **humpback grouper** has black spots that gradually increase in number as the fish grows.

 The **glass squid**, also known as the cockatoo squid, lives between 4,900 and 8,200 feet (1,500 and 2,500 m) deep and has a completely transparent body. It can light up, making its internal organs visible.

 Captain **James Cook** crossed the Indian Ocean three times: first through Indonesia, then slightly north of Antarctica, and finally south of Australia.

 The **Andaman coral reefs** are home to some of the most spectacular species of marine wildlife in the world. More than 200 species of *Madreporaria*, or stony coral, and more than 500 species of fish live there.

 The **blue-ringed octopus** is barely 8 inches (20 cm) long but is one of the most dangerous animals in the world. It carries enough venom to kill 26 people.

 There is so much **naval traffic** in the Strait of Malacca that vessels like cargo ships sometimes capsize or collide.

The **minke whale** is a small whale that only reaches about 18 feet (5.5 m) long. It is one of the most common whales, with more than 1 million living in oceans all over the world.

 The **silver hatchetfish** lives up to 1,970 feet (600 m) below sea level and its body is covered in thick, shiny scales that reflect the light of its bioluminescent organs.

 The **wandering albatross**, a seabird, has a larger wingspan than any other bird—11.5 feet (3.5 m). Its large wings allow it to glide in the air, and it can fly for many days at a time.

 The **yellow-bellied sea snake** is the most common species in the Indo-Pacific. It can grow up to 10 feet (3 m) long, and lays eggs on dry land.

 The **striped bonito** is similar to a small tuna. It can grow up to 3 feet (1 m) long and weigh 22 pounds (10 kg). It feeds on squid, fish, and crustaceans.

 The **sperm whale** can dive 7,380 feet (2,250 m) deep and stay underwater for up to two hours while it hunts giant squid.

 In 1859, Englishman **Alfred Russel Wallace** identified an imaginary boundary, the Wallace Line, in Indonesia that separates species of Asia to the west, from those of Australia to the east.

 An extraordinary shipwreck was found in the Java Sea, containing thousands of Chinese **porcelain** objects still intact after 800 years at the bottom of the sea.

 From the seventeenth century to the twentieth century, **whale oil** extracted from whale blubber was used to power lamps. Today whaling is banned in many countries, including the United States.

 After the 2004 tsunami, **warning buoys** and **warning signs** were installed in the water and on land to alert people in the event of danger.

 Excavating devices remotely controlled from ships are used to extract minerals from the bottom of the sea.

 The **proa** is a sailboat from Malaysia and Indonesia that is light on the water and moves very quickly. It is used by traders, fishermen, and even pirates.

 The **mahi-mahi** or **dolphinfish**, which can be up to 6.5 feet (2 m) long and can weigh 88 pounds (40 kg), is brightly colored, extremely quick, and can also leap high above the water.

 During World War I, German Navy captain Karl von Müller attacked 30 British ships in three months aboard the **SMS Emden**. He was eventually defeated and captured on the Cocos Islands, a territory of Australia.

 Starfish are actually not fish; they are related to sand dollars and sea urchins. The purple sea star lives in very deep waters and has a fragile, flat body.

 The **Dutch East India Company** fought the British East India Company for centuries for control of trade in products like nutmeg.

 Nyai Roro Kidul is the legendary Indonesian goddess of the sea. She is also the patron of gatherers of swallow nests, who once risked their lives to obtain this ingredient for a famous soup.

 The **bigeye tuna** is mostly caught in tropical waters. It can grow to more than 6.5 feet (2 m) long, and has a dark blue metallic back and a white belly.

 The **king of herrings**, also known as the giant oarfish, is completely unrelated to herrings. It has a long, ribbonlike body that can be up to 36 feet (11 m) long.

 Drones are used to search for shipwrecks at the bottom of the Indian Ocean.

Pacific Ocean

RUSSIA

Sea of Okhotsk

Oyashio Current

Bering Sea

Oyashio Current

Kuroshio

Japanese Sea

JAPAN

Yellow Sea

South China Sea

ASIA

Kuroshio Current

Westerly Winds

Philippine Sea

Sulu Sea

Celebes Sea

The Equator

North Equatorial Current

Equatorial Countercurrent

INDONESIA

PAPUA NEW GUINEA

Arafura Sea

East Australian Current

Coral Sea

Great Barrier Reef

Westerly Winds

Link to download Pacific Ocean maps:
www.blackdogandleventhal.com/MapsoftheWorldsOceans/Pacific_Ocean

AUSTRALIA

Tasman Sea

PACIFIC OCEAN

NEW ZEALAND

TASMANIA

Antarctic Circumpolar Current

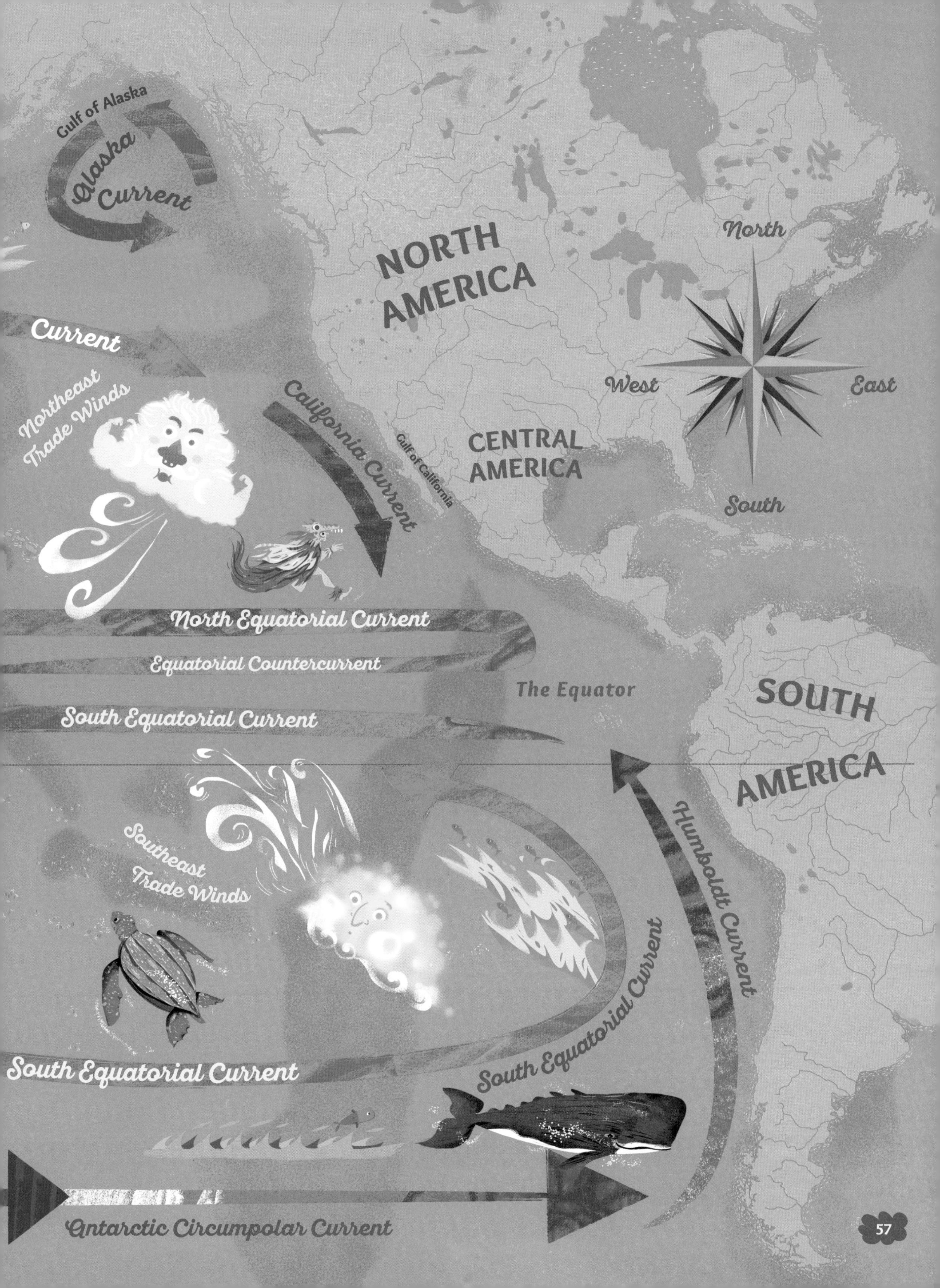

Gulf of Alaska

Alaska Current

Current

Northeast Trade Winds

California Current

Gulf of California

NORTH AMERICA

North

West

East

CENTRAL AMERICA

South

North Equatorial Current

Equatorial Countercurrent

South Equatorial Current

The Equator

SOUTH AMERICA

Southeast Trade Winds

Humboldt Current

South Equatorial Current

South Equatorial Current

Antarctic Circumpolar Current

Pacific Ocean
Infinite Routes

The Pacific Ocean is sometimes called the "aquatic continent." It is 40 times larger than Europe and has tens of thousands of islands.

Approximately 60,000 years ago, the first humans who crossed the Bering land bridge, which is thought to have become the archipelago of Indonesia after the end of the Ice Age and the rising of the seas, found themselves facing the largest ocean in the world.

Back then, they may have been able to cross short stretches of sea, but not venture out into the open ocean. However, when Indonesia became an archipelago, they learned to move among the islands, making canoes that could brave the sea.

The Pacific Ocean contains large islands like Japan, but its main regions (Melanesia, Micronesia, and Polynesia) are made up of mostly small islands with small populations. Ancient inhabitants of the Pacific Ocean often sought out new islands on which to live.

These explorer populations built large canoes made to withstand the storms of the Pacific Ocean, carrying entire families, seeds to plant in new lands, and dogs and chickens to raise. They sometimes traveled for weeks before seeing land.

The first oceanic explorers probably departed from the island of Taiwan 5,000 years ago, arriving in the Philippines and then moving eastward. They discovered thousands of islands before arriving at the most distant of all, Easter Island, perhaps 1,500 years ago.

Today the Pacific Ocean is inhabited by millions of people of different origins, from indigenous Australian and Polynesian peoples to those from European and Asian backgrounds, to Aleutians from the frozen North. They live on island–states such as Vanuatu, or on islands that are part of larger nations, such as Hawaii and French Polynesia.

The Pacific Ocean is too deep to drill for mineral resources, so the economy of the region is based on fishing and tourism, with people drawn to natural treasures such as the Great Barrier Reef and hundreds of the world's most beautiful islands.

Despite its vast size, the Pacific Ocean is threatened by overfishing, pollution, and especially by rising seawaters, which are already submerging islands like Kiribati and forcing the population to leave.

The Pacific Ocean, the largest ocean in the world, is larger than the Atlantic and Indian oceans combined. This immense water world extends from the Arctic to the Antarctic, from Asia to the Americas, and is inhabited by dozens, perhaps hundreds, or even thousands of species, from plankton to colossal blue whales.

The coldest part of the Pacific Ocean is along the sixtieth parallel south, near Antarctica. It is home to whales, seals, and penguins.

The Pacific Ocean is dotted with islands that are sometimes thousands of miles away from the mainland. There may be more than 25,000 islands in the ocean, the largest being Australia, and the smallest, tiny coral reefs inhabited only by fish and birds that nest among the coconut palms.

The Mariana Trench in the Pacific Ocean is the deepest trench in the world, and the Challenger Abyss is the deepest point of the trench. It is 35,825 feet (10,920 m) below the surface and can be reached with the Bathyscaphe *Trieste*, a deep-diving research vessel.

The Coral Triangle between Indonesia, the Philippines, and the Solomon Islands is considered the global center of marine biodiversity. More than 2,000 species of different coral fish live here.

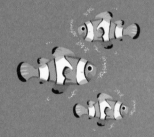

The North Equatorial Current in the Pacific Ocean is the longest ocean current in the world. It moves for 9,000 miles (14,500 km), from the Americas to Asia.

The Pacific Ocean was the last ocean to be discovered by European seafarers. It is one of the oceans marine biologists study the most, due to its biodiversity. The Galápagos Islands, off the coast of Peru in South America, is where Charles Darwin, father of the modern theory of evolution, made many of his discoveries.

The Ring of Fire is a ring of volcanoes around the Pacific Ocean. It has a horseshoe shape and is 24,850 miles (40,000 km) long. About 755 volcanoes are in the ring, and 80 percent of the world's earthquakes occur here.

Strange animals like the rare goblin shark live in the deep, dark Pacific Ocean. The goblin shark looks like a sea monster. Less than 50 of them have been caught since they were discovered in 1897.

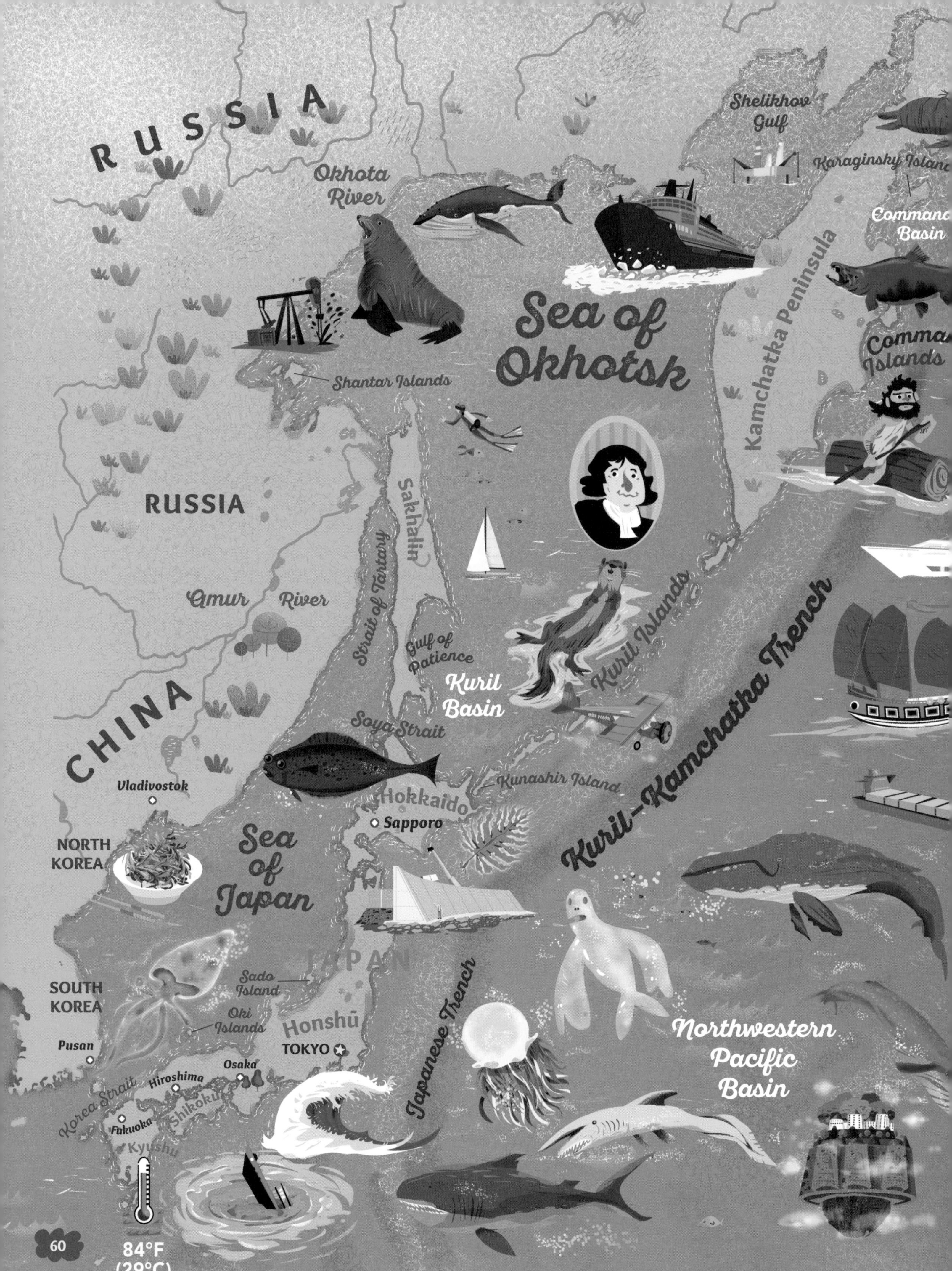

RUSSIA

Okhota River

Shantar Islands

RUSSIA

Amur River

CHINA

Vladivostok

NORTH KOREA

SOUTH KOREA

Pusan

Korea Strait

Hiroshima

Fukuoka

Kyushu

Shikoku

Osaka

Oki Islands

Sado Island

Honshū

TOKYO

JAPAN

Sea of Japan

Sapporo

Hokkaido

Kunashir Island

Soya Strait

Gulf of Patience

Strait of Tartary

Sakhalin

Sea of Okhotsk

Shelikhov Gulf

Karaginsky Island

Kamchatka Peninsula

Command Basin

Command Islands

Kuril Basin

Kuril Islands

Kuril–Kamchatka Trench

Japanese Trench

Northwestern Pacific Basin

84°F (29°C)

ALASKA

Anadyr Island

16–30°F
(-1–-9°C)

Bering Sea

Aleutian Basin

Drifting Ice Limit

Bacino Bowers

Near Islands

Aleutian Islands

Rat Islands

Andreanof Islands

Aleutian

Aleutian Islands

Trench

St. Matthew Island

Nunivak

Gulf of Alaska

Alaska Peninsula

Kodiak

Patton Seamount

Pribilof Islands

Fox Islands

Bowie Seamount

Comstock Seamount

Hawaiian–Emperor Seamount Chain

Northern

Pacific

Ocean

NORTHERN PACIFIC OCEAN

Northern Pacific Ocean

 The **walrus** is a large sea mammal, up to 12 feet (3.7 m) long and weighing up to 3,700 pounds (1,679 kg). It has large ivory tusks, which can be 3 feet (1 m) long.

 The **Japanese spider crab** is the largest-known crustacean. Its claw span can be up to 18 feet (5.5 m) and it can weigh up to 42 pounds (19 kg).

 Primitive vessels made from tree trunks or bark may have been used by the first humans who traveled from Asia to America 20,000 years ago.

 This **shamanic mask** of the Yup'ik people shows a human face emerging from the mouth of a seal, symbolizing the union between the world of the spirits and that of humans.

 This ancient **harpoon tip** made from walrus ivory looks like a squid, showing the respect that ancient fishermen in the Bering Sea had for the creatures they hunted.

 The **horned puffin** is the most colorful of the seabirds. It has a black and white body, a yellow and orange beak, and orange feet.

 Large **icebreakers** are ships that are designed to navigate through icy waters. They are often painted red to help them stand out against the ice.

 The **Alaska pollock** is a cold-water fish of the northern Pacific, where it can live at depths of up to 600 feet (180 m). It can grow to be 3.5 feet (105 cm) long and has three dorsal fins.

 The **giant octopus** has an arm span of more than 16 feet (5 m) and can weigh up to 110 pounds (50 kg).

 For centuries, Arctic peoples used the **umiak** canoe, which is made of whalebone or driftwood covered with sealskin, for transport and fishing. They are still used today, but are now made of aluminum.

 Beautiful **wooden headgear** of the Unangan people, sometimes decorated with sea lion whiskers, protected the eyes of fishermen hunting on the high seas, and also helped them hear better.

 Legend has it that about 1,500 years ago, the Buddhist missionary Hoei-Shin sailed east from China for more than 6,000 miles (9,650 km) and discovered a mythical country known as **Fusang**. Some people believe it was somewhere in the northern Pacific.

 The **mudflats of Alaska** appear at low tide and consist of masses of very fine mud that can be hundreds of feet thick at some points making them very dangerous for humans.

 The **bald eagle** is common in Alaska and has been the symbol of the United States since 1782. This predatory bird can be up to 3.5 feet (1 m) long, with a wingspan up to 7.5 feet (2.3 m).

 To avoid frightening their prey, ancient hunters on Kodiak Island hid among the rocks underneath **seal-decoy helmets**.

 The **Steller sea lion** is the largest of the eared seals. Male adults, which are larger than the females, can be more than 10 feet (3 m) long and can weigh more than 2,400 pounds (1,090 kg).

 The **Izembek lagoon** in Alaska is located at the center of a vast wildlife refuge and contains one of the world's most extensive prairies of *Zostera marina*, or eelgrass, a sea plant that has many uses.

 Modern **yachts** can travel more than 1,240 miles (2,000 km) with one load of fuel, particularly in the vast northern Pacific, where islands are rare and navigation is difficult.

 The different species of Pacific **salmon** come in different colors, like pink, red, and silver.

 Various places in the northern Pacific are named after the Danish cartographer **Vitus Bering**, who explored the region in the 1700s.

 In the cold Aleutian Islands, which extend like a bridge between Asia and the Americas, Japanese and American troops fought some of the least-known **battles of World War II**.

Halibut is a fish that can weigh more than 330 pounds (150 kg) and can seriously wound fishermen, which is why the Tlingit people created wooden hooks to catch "small" specimens weighing 33 to 44 pounds (15 to 20 kg).

 The **North Atlantic right whale** is 52 feet (16 m) long and can weigh up to 80 tons (73 metric tons). It feeds off krill, which it hunts by following their sounds.

 The wild waves offshore from eastern Japan inspired the artist Katsushika Hokusai, who depicted them in the nineteenth century, in his famous woodblock print **The Great Wave off Kanagawa**.

 This **whalebone amulet** of the Tlingit people depicts the journey of a shaman into the spirit world, lying down on the back of a two-headed sea creature.

 Deep in the North Pacific, there is a stretch of ocean about 3,730 miles (6,000 km) long and 1,240 miles (2,000 km) deep that is trapped between the water's currents and the ocean floor. Known as the **shadow zone**, this stretch of ocean is believed to have been trapped there for 2,000 years or more.

 The **goblin shark**, or elf shark, has unusual pink coloring, a floppy body, tiny eyes, a mouth equipped with slender, sharp teeth, and a head with a beaklike snout.

 The **sea otter** has webbed rear feet and smaller front feet, but it is very agile and uses its claws to collect food like sea urchins, mollusks, and crabs on the ocean floor.

 Cultivated in Japan for more than 1,000 years and consumed throughout the world today, **wakame** algae, or seaweed, is prepared in soups and salads that are very nutritious.

 The **Great Pacific garbage patch** is the largest collection of ocean plastic and other debris in the world. The trash floats between Hawaii and California.

 The **yellowfin tuna** can easily be differentiated from other tunas by its very long, bright yellow second dorsal fin and anal fin.

 The **firefly squid** lives in deep, dark waters and it uses the luminous organs strewn over its tentacles to attract the small fish on which it feeds.

 The **Naruto Strait**, between the Sea of Japan and the Pacific Ocean, is affected four times a day by strong sea currents that form whirlpools that can be up to 66 feet (20 m) in diameter.

 In 1931, a plane known as **Miss Veedol** made the first nonstop flight across the Pacific, taking off from Japan and landing in the State of Washington 41 hours later.

 Cement **seawalls** that are up to 41 feet (12.5 m) high protect almost 9,300 miles (15,000 km) of the Japanese coast, but they are not always enough for stopping tsunamis, such as the one that occurred in 2011.

 The **megamouth shark** was discovered in 1976. Since then, about 100 of them have been sighted.

 The Blob is the name given to an area of unusually warm water that has been expanding in the northern Pacific since 2013, affecting marine life and even the climate on land.

 The **Nomura's jellyfish** is a giant jellyfish that can grow up to 6.5 feet (2 m) in diameter and weigh 440 pounds (200 kg). It weighs so much that it can tear fishing nets.

 The fictional flying island of **Lugnagg** from Jonathan Swift's novel *Gulliver's Travels* is described as being 100 leagues southeast of Japan. Swift wrote that it's populated by ineffectual geniuses and their servants.

 The **Wasgo**, or **Gonakadet**, is a sea creature, half-whale and half-wolf, from northeastern Pacific mythology. Even though it is frightening-looking, it is said that seeing it brings good luck.

 Ningen are a kind of Japanese mermaid that some Japanese fishermen have claimed to see. They are large white creatures with a vaguely human form.

Ryukyu Islands

Okinawa Island

Nansei-Shoto Trench

Mariana Trench

Mariana Trough

12

Philippine Sea

Philippine Trench

Danao

West Philippine Basin

Mariana Islands

Challenger Deep

Palau

Moluccu Sea

Maluku Islands

Asia

MARIANA TRENCH

PACIFIC OCEAN

INDIAN OCEAN

Australia

Indo-Pacific Ocean

 The fishermen and sailors on the island of Taiwan address their prayers to the **sea goddess Mazu**, who may have been an actual "sorceress" who lived 1,000 years ago in the Chinese region of Fujian.

 Hydrophones were originally used to monitor the movements of military submarines. They also record sounds of the underwater world, like the calls of sea creatures.

In 1997, a hydrophone located in the Pacific recorded **the Bloop**, a sound so strong that people thought it must be related to a gigantic sea creature. It actually came from a strong earthquake in Antarctica.

 The **Indo-Pacific dolphin** is a sea mammal that can grow up to 10 feet (3 m) long and weigh up to 510 pounds (230 kg). It communicates with sounds like clicks, whistles, and vocalizations.

 The rear pair of claws of the **pelagic red crab**, or tuna crab, change into little flippers that allow it to swim. Males are blue and females are greenish in color.

 Similar to reinforced camping tents, modern **inflatable lifeboats** are indispensable for ships that cut through the stormy Philippine Sea.

 The ancient Chamorro made decoys, called **poiu**, with half a coconut and a stone ballast to lure fish near fishing boats.

 Ha Long Bay has nearly 2,000 limestone islands and islets that are up to 656 feet (200 m) high. They were created by erosion, when the bottom of the bay emerged from the water.

 The **Chinese crayfish** is just 4 inches (10 cm) long and is one of the most frequently fished species in the China Sea, where it lives between 6.5 and 655 feet (2 and 200 m) below the surface.

 According to one theory, 5,000 years ago the ancestors of the **Polynesian people** departed from the island of Taiwan, beginning one of the greatest adventures of discovery in human history.

 The **water monitor** is a reptile with a long neck, which can grow up to 6.5 feet (2 m) long. It has sturdy paws and a flat tail similar to a crocodile.

 Some mysterious disappearances of ships in the **Romblon Triangle** may have occurred due to eruptions of the region's numerous submerged volcanoes.

 The island of Hainan, located in a warm subtropical band and bathed by long waves that are ideal for **surfing**, is an international attraction for enthusiasts of this sport.

 The **horned sea star** looks like a cookie covered with chocolate chips. It has five triangular arms with dark-colored, pointed tubercles.

 Long side floats stabilize the typical **Philippine fishing boat**, which is propelled by a diesel motor.

 Ancient inhabitants of the Mariana Islands, known as the Chamorro, used **axes** carved from hard, sharp **Tridacna** (a kind of saltwater clam) shells.

 Stilt fishing, also practiced in the coastal waters of Vietnam, requires fishermen to maintain a good sense of balance. It takes about a month to learn to balance upright.

 The **pygmy seahorse** is the world's smallest seahorse. It is only up to ¾ inch (2 cm) long and can camouflage itself on sea fans, a strategy that protects it from predators.

 The **Philippine Sea** has many tidal waves because it is an area of very active volcanoes. It is carefully monitored for **tsunamis**, which rarely occur.

 A **milkfish** can grow up to 5.5 feet (1.7 m) long, but it has no teeth and can only eat algae and small invertebrates.

 According to Philippine mythology, the giant sea creature **Bakunawa** causes eclipses by swallowing up the Moon.

 Italian explorer and scholar **Antonio Pigafetta** traveled with Ferdinand Magellan, documenting his voyage around the world in journals.

 About 3,500 years ago, the Mariana Islands may have been the second stopping place for prehistoric sailors from Asia who populated the Pacific, creating the **Lapita** civilization.

 In the delta region of the Mekong River, which traverses all of Vietnam from north to south, **mangroves** create a dense forest that at some points can extend up to 24 miles (39 km) from the coast.

 The **common thresher** is one of the strangest-looking sharks. It has a long tail with a very pronounced upper lobe that makes up nearly half its body length.

 In 2009, the state of Palau created a **shark sanctuary**, forbidding commercial fishing in an area as large as France, where more than 100 different species of sharks have been identified.

 The **beaked sea snake** is difficult to see because it's light gray in color. Its bite is lethal to humans.

The *Tian Kun Hao*, a Chinese **dredger**, is 460 feet (140 m) long, and is capable of excavating material at depths of up to 115 feet (35 m) and then transporting it to where it is needed.

 The **mandarin fish** is named for its mix of green, yellow, orange, and blue coloring, reminiscent of important dignitaries in the imperial court of China.

 When the **starry toado** is in danger, it quickly swallows water, changing into a rough, spiny ball that is impossible to eat.

 The **imbricated**, or **hawksbill, sea turtle** is named for the sturdy plates that cover and partially overlap its heart-shaped shell.

 The **leopard shark** can grow to 11.5 feet (3.5 m) long, but it is harmless. As an adult it has a spotty coat, but young leopard sharks have zebra stripes.

 The Portuguese explorer **Ferdinand Magellan** was the first European to come ashore on the Philippines, in 1521. He died there that same year, suffering the same fate as many great explorers.

 When it feels threatened, the small **blue-ringed octopus** shows its colorful rings, a warning of how dangerous it might be if attacked.

 On the island of Palau, stories and legends are carved onto **illustrated storyboards** made of hardwood that are sometimes shaped like fish.

 The seafaring **Iranun** people, ancient pirates and heroes of the Philippine struggle for independence, are followers of Islam who live both in the Philippines and in Sabah, Malaysia.

 Bagangs, in Borneo, are bamboo structures used for fishing anchovies. Equipped with nets that are dropped down to the ocean floor, they are so light that they seem to float on the water.

 In 1871, the volcano Vulcan erupted and destroyed the settlements on Camiguin Island in the Philippines, including a cemetery. Today, a floating cross marks the location of the **sunken cemetery**.

 The **mangrove horseshoe crab** is more closely related to spiders and scorpions than to crabs. It can grow up to 1 foot (30 cm) long and has a long, rigid tail.

 Legend has it that an abandoned bride who gazed out at sea, waiting for her husband, was transformed into rock on **Mount Kinabalu** on the island of Borneo.

 The **Indonesian coelacanth**, discovered in 1997, is considered a living fossil, like the similar coelacanth discovered in 1938 in the waters of the Mozambique Channel.

 There are millions of small golden jellyfish in **Jellyfish Lake** on the island of Eil Malk in Palau.

 The **humphead parrotfish** has an obvious bump on its head that it uses to break up large colonies of coral, which it then crumbles with its powerful, beaked mouth.

 Inhabitants of the Marshall Islands in Micronesia made **maritime maps** out of sticks, representing currents, and shells, representing islands.

 The **dugong** is an herbivorous sea mammal. To eat, it rests its mouth on the ocean floor and tears up the roots of sea plants while grazing in underwater grasslands.

 The **sea wasp** is a very small, almost transparent jellyfish with 60 tentacles that can extend up to 10 feet (3 m). Its sting is very painful and dangerous.

 The **nautilus** is related to the squid, but it has an extremely beautiful and colorful shell, shaped like a spiral, with up to 90 tentacles protruding from the opening.

 Gadao, a legendary chief on the island of Guam in Micronesia, had such extraordinary strength that he once broke a canoe in half while rowing against an opponent.

 In 2012, the **Deepsea Challenger**, a single-seat deep-sea diving submersible, reached the deepest-known point on Earth, 364,000 feet (111,000 m) deep, with one man on board.

Mariana Trench

The Mariana Trench is the deepest depression in all the world's oceans. It is 1,580 miles (2,542 km) long and marks the boundary between the Pacific and the Philippine continental plates.

Total Darkness

-656 feet (-200 m)

-3,281 feet (-1,000 m)

-9,843 feet (-3,000 m)

-16,404 feet (-5,000 m)

...to find the deepest point of the oceans for many years. The Mariana Trench was discovered in 1875, during an expedition carried out by the HMS *Challenger*. It measured a depth of 26,850 feet (8,184 m). Today we know the trench is actually 35,755 feet (10,908 m) deep.

Humans have visited the Challenger Deep less often than the Moon. As of today, it has only been seen by three people: two members of the Bathyscaphe *Trieste* team in 1960, and the movie director James Cameron, who descended there alone in 2012, using a futuristic single-seater submarine.

Despite enormous pressure and cold, sea life is present at the bottom of the Mariana Trench.

–22,966 feet (–7,000 m)

–26,247 feet (–8,000 m)

–29,528 feet (–9,000 m)

–32,808 feet (–10,000 m)

New Britain Trench

Solomon Sea

Kiriwina Island

Goodenough Island

Ferguson Island

Normanby Island

Sideia Island

Basilaki Island

Milne Bay

Goodenough Bay

Orangerie Bay

Papua Abyssal Plain

Coral

Papua New Guinea

★ PORT MORESBY

Coral Sea Basin

Queensland Plateau

Great

Great Papuan Plateau

Gulf of Papua

PLEASE SAVE THE REEF

Torres Strait

Cape York

Cape Grenville

Cape York Peninsula

Princess Charlotte Bay

Cape Melville

Cape Melville National Park

Jack River National Park

Cape Flattery

Sea

Great
Barrier Reef
Marine Park

Barrier

Reef

AUSTRALIA

AUSTRALIA

Great Barrier Reef

Cairns

Trinity Bay

Mission Beach

Wooroonooran National Park

Hinchinbrook Island

Palm Island

Magnetic Island

Townsville

Halifax Bay

Tully Gorge National Park

Koombooloomba National Park

Burdekin River

Whitsunday Islands

Airlie Beach

Conway Beach

Cape Conway

Mackay

Long Island

Townshend Islands

Cape Clinton

Curtis Island

Flinders River

Barron River

Great Barrier Reef

QUEENSLAND

Australia

PACIFIC OCEAN

INDIAN OCEAN

Tasmania

71

Great Barrier Reef

 The warm ocean waters east of Australia help transform **shipwrecks** from World War II into wonderful gardens of coral.

 The Australian Great Barrier Reef is very important to the region. Australian **activists** are fighting to preserve it.

 The **blue stingray** is an average-sized ray that is usually found on the ocean floors of coral reefs.

 The **Australian humpback dolphin** is similar to the common dolphin, but it has a dorsal fin that forms a sort of hump.

 The **minke whale**, or lesser rorqual, can grow up to about 32 feet (10 m) long, but in the Australian waters a pygmy subspecies has been identified that does not exceed 24 feet (7.5 m).

 The **green sea turtle** is the most common turtle in tropical and subtropical waters. It can usually be found in shallower coastal areas, where it grazes on sea plants.

 To visit coral without damaging them, **catamarans** are much more suitable than "monohull" vessels because the two hulls don't sink as far into the water and are easier to maneuver.

 The **fiddler crab's** name derives from its claw, which moves like the bow of a violinist.

 The **coral trout**, despite its name, is a reddish grouper densely covered with blue flecks. It is a great hunter that can reach up to 4 feet (1.2 m) long.

 The colorful **butterflyfish** adapts well to even the most tangled coral formations, where it can swim easily thanks to its slim body.

 Staghorn coral form tangled colonies and branches that are pointed at the ends, like deer horns. They are very common, particularly in the first 65 to 98 feet (20 to 30 m) below the surface.

 Six cannons from the **HMS Endeavour**, commanded by James Cook, the European discoverer of Australia in the 1700s, were found on the reef, where they had been thrown overboard in order to lighten the ship.

 The **oceanic whitetip shark** has white patches on the tips of its dorsal fins. It can usually be found not moving on sandy ocean floors during the day.

 The **Napoleon fish** owes its name to its prominent frontal hump, which resembles the shape of the hat worn by Napoleon Bonaparte. A true giant of the reefs, this fish grows more than 6.5 feet (2 m) long.

 Table coral belongs to the Acropora genus, perhaps the most widespread species on the coral reefs. It grows predominantly in a horizontal direction, forming flat structures 6.5 to 10 feet (2 to 3 m) wide or more.

 The **mini-submarines** that take tourists to admire the coral reef are very light, less than 200 pounds (90 kg), and they can descend to depths of 147 feet (45 m).

 Vase sponges are among the most common and easy-to-recognize sponges along the Australian Great Barrier Reef.

 The **white-bellied sea eagle** is a sacred bird to the indigenous populations of Australia, and figures predominantly in mythical tales.

 The **mimic octopus** escapes predators by changing color and shape.

 The **Great Barrier Reef** is one of the most amazing spectacles in nature, with more than 400 species of hard and soft coral of every shape and color.

 The **grey reef shark** is not very large, but it is aggressive toward other shark species, pushing them away from their hunting grounds. It rarely ventures into the open ocean.

Tropical cyclones that form on the Coral Sea are very violent and cause difficult weather conditions on the east coast of Australia.

 The **turtleheaded sea snake** can grow up to 40 inches (1 m) long and eats fish eggs, which it seeks out by swimming slowly among the coral.

 The **blue hole** is a natural cavern, or cave, in the Great Barrier Reef that was formed thousands of years ago when the area was dry.

 The conditions of the coral barrier reef depend on the water, but also on the way in which the reef interacts with the air. This is one of the many complex phenomena that have been analyzed by modern **scientific buoys**.

 Gorgonians, or **sea fans**, are similar to coral and are shaped like large fans. They are usually very colorful and can grow to be 2 feet (60 cm) high.

 The **crown-of-thorns starfish** presents one of the most serious threats to coral. It moves along the ocean floor with arms covered with poisonous spines. It devours coral polyps, leaving behind a trail of dead coral.

 The **striped surgeonfish** has horizontal yellow and blue stripes and lives in the top few feet of the reef, where algae are more abundant.

 The Great Barrier Reef slows down the powerful currents of the Pacific Ocean, creating **waves** suitable for surfing.

 The **sea crocodile** is the largest living reptile and one of the most efficient predators on the Australian coasts. Up to 20 feet (6 m) long, it also pushes out into the open sea and can even kill sharks.

 The **zebra seahorse** is less than 4 inches (10 cm) long and is easily recognizable with its light-and-dark stripes and yellow-tipped spines.

 Brain coral gets its name from the rounded shape of its colonies and the numerous, tangled wrinkles that cover its surface, quite similar to those of a human brain.

 The **giant clam**, the largest of bivalve mollusks, can be up to 5 feet (1.5 m) long and can weigh more than 440 pounds (200 kg).

 More than 1,500 species of fish swim in the underwater universe of the Great Barrier Reef, some forming **schools** that help protect them from predators.

 Nudibranchs are mollusks that have lost every trace of a shell. They have a very wide variety of shapes, but all have wonderful colors, which is why they are also nicknamed "sea butterflies."

 The **moha moha** is a legendary sea creature that looked like a long turtle with a fish tail. It was supposedly spotted in 1890 on Fraser Island. It has not been seen since.

 The **Atolla jellyfish** is a master at the art of escaping its predators. If attacked, it flashes on and off like a flashlight, attracting other predators that will free it from its attacker.

 One hundred million years ago, the **ichthyosaur** was a large marine reptile that lived in the waters of Australia, which was then located much farther south than it is today, and was still joined to Antarctica.

Great Barrier Reef: A Continent of Coral

The Great Barrier Reef is the largest living organism in the world. It is 1,425 miles (2,300 km) long and home to 1,600 different species of fish.

Changes in the world's climate have affected the Reef in negative ways. In some places, half of the coral has been lost. Scientists hope that the great biodiversity of the barrier will help reduce damage.

British naval officer Captain James Cook discovered the Great Barrier Reef when his ship the *Endeavour* got stuck in the coral in 1770.

Tasman Sea

AUSTRALIA

Australia

TASMAN SEA

Tasmania

New Zealand

PACIFIC OCEAN

Sydney

CANBERRA

Melbourne

Bass Strait

Ré Island

Furneaux Islands

Launceston

TASMANIA

HOBART

Tasman Abyssal Plain

Lord Howe Island

Tasman Sea

Tasman Basin

73°F
(23°C)

SEA SALT

North Fiji Basin

Norfolk Basin

Three Kings Islands

Kermadec Islands

Kermadec Trench

AUCKLAND

Manukau

NEW ZEALAND

North Island

Hikurangi Trench

Cook Strait

Wellington

Pacific Ocean

NEW ZEALAND

South Island

Christchurch

Southern Cross

Chatham Islands

Bounty Trough

Fiordland National Park

Dunedin

50°F (10°C)

Stewart Island

Campbell Plateau

Bounty Islands

Tasman Sea

 Abel Tasman was the first European to reach and sight Tasmania and New Zealand, in 1642. A century would pass before other explorers followed.

 In the Tasman Sea, the sharp, pointed **Ball's Pyramid** is the tallest volcanic stack in the world, 1,844 feet (562 m) high. A volcanic stack is a sea rock made of lava.

 In Maori myths, the ocean and land are opposing natural kingdoms, as seen in the endless battle between **Tangaroa**, king of the sea, and his brother Tane, god of the forests.

 The **striped marlin** is a deepwater fish that swims very rapidly. It has a shape similar to a swordfish and blue vertical bands on its body.

 Many **whirlpools** form in the Tasman Sea. They are created in a way that's similar to the formation of hurricanes, but they are not destructive.

 The **common seadragon** is a symbol of the Australian state of Victoria. It is related to seahorses and has a body with leaflike limbs.

 The **little blue penguin** is the smallest type of penguin, and is named for the bluish feathers on its back, which helps it to blend into the sea.

 Poor Knights Islands Marine Reserve is an extraordinary underwater paradise, full of sea life. It is forbidden to fish within 1.25 miles (2 km) of the reserve.

 The **giant Australian cuttlefish** can grow up to 20 inches (50 cm) long, not counting its tentacles, and it can weigh up to 23 pounds (10 kg).

 In 1955, at the height of the Cold War, the British government considered the remote **Kermadec Islands** as a site for nuclear tests. The prime minister of New Zealand refused to permit the tests.

 Indigenous Australians' **nawi** canoes are simple but clever. A skilled person can build one in a single day, using eucalyptus bark hardened over a fire.

 The **Tasmanian giant crab** can weigh more than 39 pounds (17 kg). In the males, one of its two claws is oversized, reaching up to 16 inches (40 cm) long.

 Hydrothermal volcanic chimneys have been discovered off the coasts of New Zealand, along with the submerged Tonga-Kermadec Ridge, which is full of underwater volcanoes.

 The **carpet shark** is named for the flat shape of its body and its wide head, which has appendages on the sides that look like the fringe of a carpet.

 Ancient maps suggest that **Portuguese ships** first caught sight of Australia in the early 1500s.

 Egyptian hieroglyphs discovered in New South Wales recount an ancient shipwreck. They may have been sculpted by Englishmen who had fought in Egypt against Napoleon Bonaparte.

 The island of **Motu Kokako**, now Piercy Island, is said to be the landing place for one of the canoes that brought the Maori people to New Zealand 700 years ago.

 The North Island of New Zealand resembles a fish when viewed from above. According to legends it was fished by **Maui**, the Hercules of Polynesia, thanks to a magic hook.

 The **red abalone** is a large, edible sea snail with a reddish brown exterior shell and a beautiful, silvery interior shell.

 The **hei matau** is an important cultural symbol for the Maori. Shaped like a fishhook, it is traditionally made of bone or greenstone and represents abundance, fertility, and safety at sea.

 An unidentified body of a fish came ashore on a beach in Tasmania in 1962 and caused a sensation. It was 20 feet (6 m) long, furry, and smelly, and called a **"sea monster."** Eventually it was identified as whale blubber.

The **hairy brachionus** is a strange fish, with pectoral fins transformed into organs suitable for walking on the ocean floor, and extremities similar to hands.

 The **Southern Cross** is the name of the three-engine plane that carried out the first flight from Australia to the United States, 7,250 miles (11,670 km), guided by the Southern Cross constellation.

 The **red velvetfish** is related to the scorpionfish, but it has a soft, velvety skin studded with small growths.

 The **Southern bluefin tuna** is similar to the Mediterranean red tuna. It migrates from the waters of Indonesia to Tasmania and then back again.

 The **orange roughy** can live at depths of 5,910 feet (1,800 m) and can live up to 150 years.

 Messages in bottles that were cast out to sea for various reasons can cross the oceans, helping oceanographers to better understand the sea currents.

 The **Port Jackson shark** is named for the famous Port Jackson Harbor in Sydney, Australia. It has small, pointed teeth that it uses to crush prey like sea urchins, mollusks, and crustaceans.

 In Tasmania **salmon breeding** has been practiced for more than 30 years. It limits the negative effects of fishing in the open sea, but some people criticize the practice for causing pollution in coastal waters.

 The waters of Tasmania are clean and rich in nutrients like **salt**.

 In the mid-twentieth century, air transports for New Zealand were made possible by large **seaplanes** based on the ones used in World War II.

 The **ox-eyed oreo** fish lives at depths of 720 to 5,090 feet (220 to 1,550 m) below the surface. It is not named after the cookie. Its scientific name is *Oreosoma atlanticum*, meaning "mountain body."

 Around 80 different species of sharks can be found in the waters around New Zealand, including the **Great white shark**.

 The **giant oarfish** is a ribbon-shaped fish that can be more than 32 feet (10 m) long; because of its shape, it could be the origin of the legend about sea serpents.

 Leatherback sea turtles, true giants among turtles, gather along the coasts of New Zealand to feed before migrating toward warmer waters.

 The Maori were fearsome warriors. Their magnificently painted war canoes, called **waka taua**, were carved from a single tree trunk and could be up to 130 feet (40 m) long.

 In the nineteenth century, thousands of prisoners were sent from Europe to the most distant colonies, traveling for months on terrible **prison ships**.

 Geological traces and ancient Maori legends suggest that 500 years ago a **meteorite** caused devastating tsunamis in the Tasman Sea, but not all scholars agree.

 The **leopard seal** is the third largest seal in the world. They can grow up to 10 feet (3 m) long and weigh up to 1,300 pounds (590 kg).

 It took two rowers exactly two months to cross the Tasman Sea in 2008 in a custom kayak called **Lot 41**.

 The **beach of Moeraki** on New Zealand's South Island has large boulders on its shore that are nearly 6 million years old.

 Doubtful Sound is one of the most beautiful fjords in New Zealand. A fjord is a long, narrow body of water surrounded by cliffs.

 The **longnose sawshark** uses its long snout both to hunt down prey hidden beneath the sand, and as a weapon to defend itself when attacked.

Polynesia

Tuamotu Islands

Tiputa
Mataiva · Ohe · Manihi · Takaroa
Tikehau · Rangiroa · Arutua · Takapoto
Makatea · Aratika · Apataki · Kauehi · Raraka
Kaukura · Taiaro · Tuanake
Toau · Katiu · Tae
Niau · Fakarava · Tepoto · Maka
Faaite · Tahanea · Hiti
Onaa · Motutunga · Haraik
Reito

Society Islands

Motu One
Maupihaa · Tupai
Manuae · Maupiti · Tahaa · Huahine
Bora Bora · Tetiaroa
Uturoa · Raiatea
Maiao · Moorea · Tahiti
PAPEETE
Mehetia

Leeward Islands

Windward Islands

Hereheretu

Anuanuraro
Anuanurung
Nukute

Austral Islands

Maria
77°F
(25°C)
Rimatara
Rurutu
Mataura
Tubuai
Raivavae

Eiao
Hatutu
Ua Huka
Motu Iti
Nuku Hiva
Taiohae
Ua Pou
Hiva Oa
Tahuata
Mohotani
Fatu Hiva

Marquesas Islands

Tiki
Basin

North
America
PACIFIC
POLYNESIAN
ISLANDS
Australia
New
Zealand
OCEAN

Pacific
Ocean

Tepoto
Napuka

Pukapuka

86°F
(30°C)

akume
Fangatau
Raroia
Fakahina
Nihiru
Rekareka
okota
Tauere
Tatakoto
Tupapati
Amanu
Hikueru
Hao
Marokau
Pukarua
Nengonengo
Paraoa
Akiaki
Vahitahi
Reao
Manuhangi
Vairaatea
Nukutavake
Ahunui
Pinaki

Gambier Islands

Vanavana
Tureia
Tenarunga
Tenararo
Matureivavao
Moruroa
Vahanga
Marutea
Tematangi
Maria Est
Fangataufa

Mangareva
Rikitea
Morane
Akamaru
Taravai
Temoe

0
01010
101010101

81

Polynesia

 The **breadfruit tree** can grow up to 85 feet (26 m) tall. Its fruit is similar to a small melon, and it has a starchy pulp.

 The sailing ship **County of Roxburgh** was washed ashore by a typhoon in 1906. It slowly rusted on the Atoll of Takaroa and is now a well-preserved wreck.

 The French painter **Paul Gauguin** came to the Polynesian Islands in the late 1800s where he created some of his most famous paintings.

 The **skipjack tuna** is plentiful in the Pacific and it is caught using sustainable techniques like rods and lines.

 The **Sardinella** of the Marquesas Islands plays an important role in the food chain of this part of the Pacific and is also useful to fishermen, who use it as bait for catching tuna.

 The **green sea turtle** can be up to 5 feet (1.5 m) long and can weigh more than 420 pounds (190 kg). For centuries it was fished for its meat, but is now a protected species almost everywhere.

 Stone fishing is an ancient Polynesian technique where fishermen beat the water, pushing the frightened fish into a trap. It is still practiced sometimes as a ritual during celebrations.

 Legend has it that the unfortunate **Prince of Eels**, rejected by his fiancée, Hina, turned himself into a coconut.

 The **Pemphis acidula**, known as miki-miki in Polynesia, produces a very hard, sturdy wood that is used for building the islands' traditional vessels.

 Grey reef sharks are some of the most commonly seen sharks near the lagoons of Polynesia.

 Rangiroa, the largest atoll in the Tuamotu Islands, is also one of the largest atolls in the world. It is made of approximately 415 islands and islets, and its lagoon has a surface area of about 560 square miles (1,450 km²).

 It is said that **Paahonu vahine**, the beautiful goddess of the island of Raiatea, was abandoned by her fiancé and fled on a shark to avoid boredom.

 Yachts equipped for overnight stays known as **liveaboards** allow divers to visit the most remote underwater sites.

 Fakarava groupers gather every July in the Fakarava lagoon on the Tuamotu Islands to reproduce when the sea currents are strongest and can carry their larvae into the open sea.

 Black Tahitian pearls, cultivated since the 1960s and very important for the local economy, are actually raised throughout French Polynesia, not just in Tahiti.

 The **Chilean devil ray**, despite its name, also lives in the waters of Polynesia.

 Giant jellyfish were claimed to have been sighted on the island of Teahupo'o in 2010, but the report turned out to be false.

 The **great hammerhead shark** has a distinctive hammer-shaped head and a sickle-shaped dorsal fin. It can venture more than 260 feet (80 m) below the surface of the water, and enjoys eating rays.

 The **rough-toothed dolphin** can grow up to 9 feet (2.7 m) long and can weigh up to 330 pounds (150 kg). It can often be found off the coast of the Society Islands.

 The **black marlin** is a large fish that can grow to 15 feet (4.5 m) long and weighs up to 1,650 pounds (750 kg). It feeds on squid and fish, including small tuna.

 The **Pacific white-sided dolphin** grows up to 8 feet (2.5 m) long and weighs about 440 pounds (200 kg). It lives in schools made up of hundreds of different organisms and it willingly mixes with other dolphins of equal size.

The **clownfish** lives amid the tentacles of its preferred anemone. The clownfish defends the anemone, and the anemone defends the fish with its stinging tentacles.

 The government of French Polynesia is working to create a **floating city** with houses, public buildings, and shops for a community of around 300 people.

 The **crown-of-thorns starfish** is more than 18 inches (45 cm) wide and has up to 21 arms. It eats coral polyps, leaving behind only a white skeleton.

 The **red-footed booby** is a diving seabird. On land it moves awkwardly on its webbed feet, but in the air and the water it is very agile and good at catching fish.

 The **blacktip reef shark** is common along all the reefs of Polynesia and is easily recognizable by its brown coloring and black-tipped fins. It can grow up to 6 feet (1.8 m) long.

 Every year around July, **humpback whales** start to arrive in the waters off the island of Rurutu, where the males seek companions and the females give birth to their young.

 The Polynesian canoe, called a **va'a**, may have been formed from a single tree trunk connected to an outrigger, or from planks connected with **sennit**, a rope made from extremely strong plants.

 The **black sea cucumber** is shaped like a large, flat black sausage and is related to starfish and sea urchins.

 Fish and other sea life can dry quickly in the open air in hot places like Polynesia in a process called **desiccation**.

 Tiki statues in volcanic stone depict human forms and are believed to carry spirits. They have also inspired the Polynesian design style.

 After sailing halfway around the world, the French admiral **Louis-Antoine de Bougainville** visited French Polynesia in 1764 and fell in love with the place, describing it as an "earthly paradise."

 From 1960 to 1996, certain parts of Polynesia and other areas of the Pacific Ocean were sites of **nuclear experiments**, a practice that has now been stopped.

 The **leatherback sea turtle** is the largest turtle, at about 7 feet (2 m) long and weighing 1,540 pounds (700 kg). Its shell is made of a thick skin similar to leather.

 Tourist boats in brilliant colors are used by some of the nearly 200,000 tourists that visit Polynesia every year.

 A very advanced **calculation** system was invented centuries ago on the isolated island of Mangareva, which today is inhabited by about 1,200 people.

 The **lemon shark** of the Indo-Pacific has a short, sturdy body that is usually yellowish brown or gray. It generally lives in fairly shallow waters.

 The splendid Polynesian **double-hulled canoes** were up to 130 feet (40 m) long and could transport 200 people, including 100 rowers.

 The **ancient gods and spirits** of the Gambier Islands, which represented war and fertility, were depicted in wooden statues. Only a few of the originals exist today.

 At tourist sites, long **wooden walkways** allow visitors to pass over waters abundant with sea life, without damaging the area or harming themselves on the sharp coral.

 Polynesia is the kingdom of coral and a paradise for divers, who reach diving sites on small, flat-bottomed boats called **dinghies**, which cause minimal damage to reefs.

 The **Kogia sima**, or dwarf sperm whale, is less than 9 feet (2.7 m) long and weighs about 550 pounds (250 kg), making it a small-sized whale.

Middle Bank

Kauai
Kapaa
Kekaha
Lihue

Kaulakahi Channel

Kaiwi Channel

Puuwai

Niihau

Kaula

Kaiwi Channel

81°F
(27°C)

Oahu
Pearl City
HONOLULU
Kailua

Kaiwi Chan

Hawaiian Islands

Necker Island

Pacific Ocean

Asia
North America
PACIFIC
HAWAIIAN ISLANDS
Australia
OCEAN

HAWAII

Hawaiian Islands

Molokai

Kaunakakai

Lahaina

Wailuku

Wailua

Maui

Hana

Kaupo

Lanai

Kahoolawe

'Alenuihāhā Channel

Hawaiian Islands

Hawi

Honokaa

Puako

Hawaii

Kailua

Captain Cook

Hilo

Pahoa

Kilauea Volcano

Mauna Loa Volcano

Pahala

Naalehu

Molokai Fracture Zone

72°F (22°C)

Haleiwa Trench

Hawaiian Trough

Hawaiian Islands

 The **French Frigate shoals** are the remains of ancient volcanic islands located northwest of Hawaii. They are a sanctuary for underwater life hundreds of miles away from the nearest land.

 The **kingfish** is blue and green with silvery patches and red fins. It is an extremely skilled hunter of squid.

 Kite surfing is very popular in Hawaii thanks to the islands' strong ocean winds.

 The **red-tailed tropicbird** is an easily recognizable seabird because of its pure white feathers, red beak, and central tail feathers, which can be twice as long as its body.

 Cast nets make it possible to fish from land and to capture fish that are not attracted by bait.

 **Kanaloa** is an ancient Hawaiian god with the head of an octopus or squid. It is also the name of an extinct volcano in Hawaii.

 The **Laysan albatross**, or **moli** in the Hawaiian language, has a distinctive black mask around its eyes and beak, pale pink beak and feet, and a wingspan up to 6.5 feet (2 m).

 The **olive ridley sea turtle** is considered the smallest of all sea turtles, but it can dive down to 500 feet (152 m) below the surface and can travel up to 1,240 miles (2,000 km) from the coast.

 The **Hawaiian-Emperor seamount chain** is formed by a chain of underwater islands and mountains and includes approximately 80 volcanoes, most of which are 985 feet (300 m) below the surface of the water.

 Alaea salt produced on the Hawaiian islands is reddish in color thanks to the volcanic clay with which it is mixed. It is used for cooking, cleaning, and purifying.

 Ancient surfers from 400 years ago seem to be depicted in **petroglyphs**, which are rocks with designs carved on them, on the Island of Oahu.

 In certain areas, **manta rays** swim close to the lights along the coast that lure large numbers of plankton shrimp, which these large fish greatly enjoy.

 The **tiger shark** is named for its striped body and is considered the garbage collector of the seas. Bags of trash and even automobile license plates have been found in its stomach.

 The **USS Independence** is one of the most advanced military ships that crosses the waters to Hawaii. It is fast, almost undetectable to radar, and designed to operate in coastal waters.

 The **melon-headed whale** is one of the most frequently sighted cetaceans in Hawaii, where it forms schools made up of as many as 1,000 specimens.

 The **long-beaked common dolphin** is known for its prominent and pointed beak, its triangular dorsal fin, and its incredible ability to make acrobatic leaps outside the water.

 The Hawaiian **halophyte** is a sea plant with light green roots that is much sought-after by sea turtles.

 The ancient oceanic art of paddle surfing evolved into **stand-up paddleboarding**, a sport celebrated by a championship held every year between the islands of Molokai and Oahu.

 Another ancient technique used in the Hawaiian islands is **harpoon fishing**. It is a simple technique that makes it possible to fish even from cliffs without a canoe.

 The battleship **USS Arizona** sank in 1941 in Pearl Harbor after it was attacked by the Japanese army, killing the 1,177 sailors on board. Today, there is a memorial above the sunken ship that can be reached by boat.

 The hard "sword" (known as a beak) of the marlin and shark teeth were once two components of the sharp **weapons of Hawaiian warriors**, such as spears and daggers.

 In December, approximately 10,000 **humpback whales** migrate from Alaska in search of warm waters, where they can give birth to their young.

 Falls of Clyde is a 140-year-old oil tanker. It is the last four-masted sailboat in the world with an iron hull. It is anchored in the port of Honolulu.

 The ancient **Hawaiian outrigger canoe** housed entire families for weeks at a time during voyages to colonize the Pacific.

The **underwater scooter** allows tourists to view colorful coral fish, sea turtles, and more underwater without the training needed for scuba diving.

 The **pandan** is one of the islands' most useful plants. The flesh from its fruit can be used to make excellent flour and flavoring, and its leaves can act as containers for cooking food.

 The outrigger canoe is a historic vessel that brought the first inhabitants of Hawaii from Polynesia. Today, it is used in important **international competitions**.

 The crescent shape of the ancient volcanic crater of **Molokini**, which is south of Maui, provides a perfect natural shelter for yachts traveling to Hawaii.

 The **Aegir** was an underwater human habitat that was tested in Hawaii in the 1970s. It was named after the Norse mythology sea god and is no longer being used.

 The **coconut palm** is a symbol of the tropical islands. Its trunk can be up to 98 feet (30 m) tall and 20 to 27 inches (50 to 70 cm) wide at the base.

 The **Hawaiian monk seal**, like those of the Mediterranean, is one of the rarest sea mammals in the world. It is considered an endangered species and is protected on all the islands.

 The Hawaiian islands are famous for **deep-sea fishing**. In 1954, a marlin weighing 1,000 pounds (454 kg) was caught there with the help of a decoy built by a local skipper.

 The **lava flow** on the Big Island of Hawaii gushes continually from the Kilauea volcano and ends in the ocean as a scorching hot river, causing the waters to boil.

 The **Pseudorca**, or false killer whale, is particularly numerous and observed in Hawaii. Up to 20 feet (6 m) long and weighing up to 2.2 tons (2 metric tons), these cetaceans can also feed off sharks.

 Today's **trimarans** are so light and swift that when the wind is strong, they risk getting caught up in a wave and capsizing.

 The **sailfish** has a beak like a swordfish and a large, tall dorsal fin that stays folded up in a cavity on its back when it swims at top speed.

 The University of Hawaii is carrying out experiments in the new field of **autonomous underwater vehicles (AUV)**, which are programmable underwater robots.

 The Hawaiian islands became popular with **tourists** after World War II.

 The **black sand beach of Punalu'u** is located on the slopes of the Mauna Loa volcano and is formed from black basalt crumbled by the waves. Green turtles come here to warm themselves in the sun.

 The **moonfish**, or opah, can be 6.6 feet (2 m) long and weigh 600 pounds (270 kg).

 Whale sharks that swim in the waters of Hawaii probably have traveled very long distances. In fact, little is known about their migrations.

 According to ancient legends, **Hawai'iloa**, a Polynesian hero, fisherman, and sailor, discovered the Hawaiian islands. Recent discoveries show that the archipelago was settled by humans approximately 1,700 years ago.

 The **nene** is the symbolic bird of Hawaii. It prefers prairie habitats and bushy areas (and even golf courses), where it nests among the shrubbery.

 James Cook and his crew were the first Europeans who landed in Hawaii, on the island of Kauai, in 1778.

 Mahi-mahi has beautiful blue coloring on its back and yellow coloring on its sides and belly.

 Mauna Loa is the most majestic volcano in the world. Its base is located on the ocean floor and it measures 30,085 feet (9,170 m) from base to summit. It is even taller than Mount Everest, which is 29,029 feet (8,848 m) tall.

Baja California

MEXICO

UNITED STATES

North America

Central America

SOUTHERN CALIFORNIA

PACIFIC OCEAN

Yaqui River

Ciudad Obregón

Guaymas Basin

Gulf of California

Hermosillo

Sonora River

Tiburón Island

San Esteban Island

San Lorenzo Island

The Valley of the Cirios

El Vizcaíno Biosphere Reserve

Punta Eugenia

Sebastián Vizcaíno Bay

Colorado River

Mexicali

Great Salt Lake

Colorado River Delta

Delan Basin

BAJA CALIFORNIA

Punta San Antonio

San Quintín Basin

Punta Colonet

Ensenada

San Diego

Tijuana

Punta Banda

Cedros Island

Isla Natividad

Archangel Island

Tequila XXX

Fuerte River

Los Mochis

81°F (27°C)

San José Island

Jacques Cousteau Island

Pescadero Fault

Espíritu Santo Bay

La Paz

Cabo San Lucas

Isla del Carmen

Santa Catalina Island

BAJA CALIFORNIA SUR

La Paz Bay

Santa Margarita Island

Cedros Trench

Pacific Ocean

Ballenas Bay

89

Baja California

The **California spiny lobster** lives up to 213 feet (65 m) below the surface of the water. During the day it remains hidden to avoid the octopuses and large fish that prey on it, and then emerges at night to hunt.

Dense fog, created by the cold ocean currents that condense the air's humidity, often covers the Baja California coast.

The **bullhead shark** has a large head with two crests similar to small horns. It has strong, grindstone-shaped teeth it uses to crush tough prey like sea urchins and mollusks.

The Coronado Islands were a hideout for smugglers, who transported **alcohol** to the United States when it was banned in the 1920s.

In 1539, **Francisco de Ulloa** explored the entire Baja California coast. It is said that one of his ships was tossed by a tsunami into the middle of the desert, the subject of the legend of the Lost Ship of the Desert.

Spanish explorers in the sixteenth century thought that California was an island, and for 200 years, it was drawn that way in **maps**.

The **elephant seal** is named for the large trunk that males use to make strong barking sounds, almost like roars, when they fight among themselves to win over females.

Sea lions are very friendly sea mammals. They are agile on land and can swim extremely fast, up to 34 miles per hour (55 kph).

The **Garibaldi fish** is named for its red coloring. In Italy, General Giuseppe Garibaldi's followers often wore red shirts. Similar to the Mediterranean damselfish, it measures up to 15 inches (38 cm) long and has territorial tendencies.

Today, the Baja California peninsula is a desert, but it was once covered by the sea. Shark teeth, coral, and shell **fossils** can still be found there.

The **San Ignacio Lagoon** is included in the El Vizcaíno Biosphere Reserve, the largest protected region in Latin America. In 1993 this lagoon was declared a World Heritage site.

A special **reinforced wetsuit** is required to observe the fearsome and dangerous Humboldt squid that lives in the Sea of Cortez.

In 2015, the Mexican Navy donated the old navy ship *Uribe 121* to be sunk off the coast of Baja California to create the first **artificial reef** of the North Pacific.

Divers can view great white sharks in **protective cages** a few miles off the coast of Baja.

In 1889, a gold rush broke out in Baja California, and new **steamship** lines were immediately organized in the United States to transport thousands of gold diggers there.

The **Goliath grouper** is a true giant among fish, and can grow up to 8 feet (2.5 m) long and weigh up to 800 pounds (363 kg). It is a fearsome adversary, even for small sharks.

California **kelp** can grow 2 feet (60 cm) a day. It can reach up to 230 feet (70 m) long and creates a kind of underwater forest.

The **shovelnose guitarfish** is related to stingrays. During the day, it remains hidden in the sand, but at night, it hunts crabs, worms, bivalve mollusks, and small fish.

The **angel shark** has a flat body and wide pectoral fins that allow it to rest on the ocean floor like a stingray. It can grow to 5 feet (1.5 m) in length and weigh about 65 pounds (30 kg).

The **yellowfin croaker** has a shimmering, bluish coat with bronze reflections and bands on its sides and back. It can communicate with other croakers by making drumlike sounds.

Chalchiuhtlicue is the Aztec goddess of the seas and all waters, and protector of women and children. Her name means "Woman of the Jade Skirt."

The *vaquita*, a type of porpoise, is believed to be the most endangered sea mammal in the world. It is estimated that fewer than 20 remain in the Gulf of California.

An incredible cliff that eroded near *Balandra* collapsed into the sea due to excess tourism, but it was cemented back in place.

Fish farms in Baja California are laboratories used for testing equipment and technologies used in the field of *aquaculture*.

Gray whales make yearly long migrations from the frozen waters of the Arctic to Baja California, a journey of about 6,200 miles (9,975 km) that can take two to three months.

Fish and other animals were painted on cave walls by the *Cochimi*, aboriginal inhabitants of the central part of the Baja California peninsula, 3,500 years ago.

Stretching between the ocean and the continent, Baja California is exposed to strong winds that make it ideal for *kiteboarding*. It is the site of an international championship for the sport.

The El Vizcaíno Biosphere Reserve was named for the adventurous Spaniard *Sebastián Vizcaíno*, who explored the area in the late sixteenth century.

Whale sharks reach the Sea of Cortez between early winter and late spring to take advantage of the returning currents, which encourage the growth of plankton that they eat.

The *black jellyfish*, or black sea nettle, has an umbrella 3 feet (1 m) in diameter and tentacles that extend up to 20 feet (6 m). It does sting, but is not life-threatening to people.

Shards of Ming porcelain marked the point on the coast where the *San Felipe* galleon was shipwrecked in 1576. The area was so inhospitable that the entire crew died.

The city of La Paz is home to a *bronze statue* of a mermaid playing with a dolphin.

The *common dolphin* forms schools, or groups, that grow into hundreds when there are large schools of sardines and anchovies nearby.

The Sea of Cortez, which Jacques Cousteau called "the world's aquarium," was also celebrated by *John Steinbeck*, who traveled there in 1940 and later published the book *The Log from the Sea of Cortez* in 1951.

The warm, salty, and dense air of Baja California can corrode the varnish on vessels like *fishing boats*, making them rust quickly.

The *blue shark*, which is bluish in color, is about 10 feet (3 m) long and weighs up to 400 pounds (180 kg). Females can give birth to up to 135 pups in each litter.

As legend would have it, the Mexican god *Quetzalcoatl*, who was cast out by shamans who were jealous of his power, fled by sea on a raft of serpents, swearing he would return one day.

The *Humboldt squid* is about 5 feet (1.5 m) long and has large eyes that allow it to hunt in the dark.

The *smooth hammerhead* is one of the largest hammerhead sharks, growing up to 16 feet (5 m) long and weighing up to 880 pounds (400 kg).

The *San Carlos* was one of the sailing ships that further explored Baja California in the eighteenth century.

In 1908, the American *Great White Fleet* had a layover in Baja California. This fleet of old warships was painted white and toured the world on "diplomatic visits" to various nations.

The *orca* is one of the predators at the top of the marine food chain. Even though they are known as "killer whales," they are actually dolphins.

Baja California is an ideal place for the *Great white shark* to reproduce, thanks to the abundance of prey for both adult and newborn members of the species.

The *Lagenorhynchus obliquidens*, or Pacific white-sided dolphin, lives in the northern Pacific and gets its name from its white coloration on its sides and belly.

The *blue whale*, the largest animal on earth today, swims to the waters of Baja California and the Sea of Cortez every year between February and March to reproduce.

Colón Ridge

Pinta

Pinta Channel

Marchena

Marchena Channel

Wolf
Volcano
5,600 feet
(1,707 m)

The Equator

Darwin
Volcano
4,364 feet
(1,330 m)

**San
Salvador**

San Salvador Channel

Seymour

BIENVENIDOS — WELCOME
ESTACION CIENTIFICA
CHARLES DARWIN
Charles Darwin Research Station

4,843 feet
(1,476 m)

La Cumbre
Volcano

Fernandina

3,707 feet
(1,130 m)

Alcedo
Volcano

*Cartago
Bay*

Baltra

Isabela
Bay

Pinzón

Pinzón Channel

Santa Cruz

Puerto
Isidro
Ayora

72°F (22°C)

3,688 feet
(1,124 m)

Sierra Negra
Volcano

Isabela

Puerto
Villamil

Tortuga

5,381 feet
(1,640 m)

Cerro Azul
Volcano

Puerto
Velasco
Ibarra
Volcano

Floreana

Pacific Ocean

Genovesa

Cocos Island

PACIFIC OCEAN

GALÁPAGOS ISLANDS

South America

79°F (26°C)

Galápagos Islands

Carnegie Ridge

Santa Fé

Santa Fé Channel

Esteban Bay

Hobbs Bay

San Cristóbal

Rosa Blanca Bay

PUERTO BAQUERIZO MORENO

Gardner

Española

The Galápagos Islands

 Scalloped hammerhead sharks are drawn to the most remote islands of the Galápagos, where the combination of currents of different temperatures creates underwater environments that are full of sea life.

 The **yellow stingray** is a diamond-shaped fish with large, curved pectoral fins. It has a long spine on its tail that it uses to defend itself.

 The **fur seal** has a thick and soft coat that protects it from cold water and, in the past, made it a sought-after prey for fur trappers. Today it is a protected species.

 Charles Darwin sailed to the Galápagos on the ship the **HMS Beagle**, where he studied the natural environment of the islands that had been untouched by humans for millions of years.

 The **frigatebird** is a seabird that can be easily identified by a large sac beneath its throat, which swells up during courtship, highlighting its bright red color.

 Today, it takes only a few hours to **fly** to the Galápagos, which is 560 miles (900 km) from land. In 1835, Darwin's ship took eight days to arrive there from Lima, Peru.

 The **Blainville's beaked whale** is a strong, heavy cetacean that can weigh up to 1 ton (0.9 metric ton). The males have a bulge near the corner of their mouths, which forms a long beak.

 The **pilot whale** has short, crescent-shaped side fins and one dorsal fin on the forward portion of its back. It can be up to 21 feet (6.5 m) long and can weigh 3 tons (2.7 metric tons).

 The **Galápagos shark** doesn't only live in the Galápagos. It is a reef shark that can be found in tropical seas all around the world.

 The **beaked whale** has a torpedo-shaped body and a short mouth that looks like it's smiling. It can grow longer than 23 feet (7 m) and can weigh up to 6 tons (5.5 metric tons).

 In Kurt Vonnegut's science-fiction novel *Galápagos*, shipwrecked humans transform into **furry sea creatures** as they evolve over a million years.

 Darwin's, or **Galápagos, finches**, named in honor of this great scientist, were studied by Darwin to prove that evolution helped animals adapt to different environments on each island of the Galápagos.

 For many years, fishing boats with **surrounding nets** have threatened the unique fauna of the Galápagos. This fishing technique is now forbidden in the waters of the archipelago.

 Small **Galápagos penguins**, which weigh about 5.5 pounds (2.5 kg) and are 19 inches (49 cm) tall, are able to live on these islands close to the equator thanks to the Humboldt Current and cool water from the Cromwell Current.

 The **Galápagos hawk** flies in small groups, surveying the ground in search of prey like lizards, snakes, and small rodents.

 The **red-lipped batfish** has pectoral and pelvic fins that look like paws. It uses them to make small jumps along the ocean floor.

 The **Galápagos tortoise** is the largest land tortoise on the planet. It can weigh up to 880 pounds (400 kg) and reach 6 feet (183 cm) long.

 The **Charles Darwin Research Station** was dedicated in 1964 and is entrusted with the study and preservation of the environment of the Galápagos.

The underwater robot **Deep Rover** gathers rock samples and helps draw maps of the ocean floor.

 The **blue-footed booby** spends a lot of time fishing. Its compact and waterproof feathers allow it to constantly plunge into water.

 Sea lions are often seen sunning themselves on deck chairs on the islands. Females weigh up to 220 pounds (100 kg), while males can weigh up to 550 pounds (250 kg) and grow up to 8 feet (2.5 m) long.

 Santa Cruz, the administrative center of the Galápagos, was once known as **Indefatigable**, from the name of a famous English sailing ship that had participated in the Napoleonic wars.

 The **marine iguana**, which can be up to 53 inches (1.3 m) long, is the only lizard that feeds exclusively in the sea as an adult. It can plunge 33 feet (10 m) below the surface to feed on submerged algae.

 The Galápagos were once a base for **whaling ships**, which killed thousands of whales and turtles and caused other disasters, like the fire that devastated Floreana Island in 1820.

 The islands of the Galápagos are volcanic in origin and rise up in a **hotspot**, where the magma that constantly pours out from the Earth's interior creates a new, solid crust.

 The **red rock crab**, named for its bright red coloring, is very quick and agile. It mostly eats algae, and uses its sturdy claws to cling to rocks to stop it from being carried away by the strong ocean waves.

 The geographer **Abraham Ortelius** included the islands in his world map in 1570, calling them "Galopagos" or "Galopegos."

 In 1535, **Tomás de Berlanga**, a bishop of Panama, discovered the Galápagos by chance when his ship went off course while traveling to Peru.

 Before Berlanga's arrival there, it is believed that the Galápagos were uninhabited by people. But an ancient **Peruvian flute** found on one of the islands indicated that people may have been on land at some point before 1535.

 Saildrones are aquatic robots that collect and send data to scientists who study the oceans.

 In 1704, Scottish Navy officer **Alexander Selkirk** was marooned on the Juan Fernández Islands in the South Pacific Ocean for four years and four months. He was rescued by a ship that stopped near the island on its return trip to England in 1709.

 The **Atlantic spotted dolphin** has dark and light spots over its gray body. It can grow up to 7.5 feet (2.2 m) long and weigh about 290 pounds (130 kg).

 On the island of Floreana in an area now called Post Office Bay, there is an old **barrel** that sailors from whaling ships used to drop off letters they wanted sent home. Sailors returning home would pick them up and deliver them, sometimes years later.

 The **Galápagos cormorant**, also known as the flightless cormorant, is the largest cormorant and the only one that cannot fly. It is, however, an extremely skillful and rapid swimmer, even underwater.

 Snorkeling has little impact on the environment and is an ideal way to admire the sea species of the Galápagos.

 On the island of Española, the strong sea current creates a **blowhole**, a jet of water similar to a geyser, which can spray up to 100 feet (30 m) high.

 The **Galápagos land iguana** resembles a small dinosaur, with a sturdy head, long tail, and spiny crest on its back. It feeds mostly on cactus and fruit.

 Since the Galápagos are located along routes between America and Asia, they provided a perfect hideout for **pirates** and privateers who robbed ships crossing the waters.

 Experiential fishing allows tourists to participate in local fishermen's outings. It also helps support fishermen whose commercial activity is limited in these protected waters.

Northeast Pacific Ocean

CANADA

UNITED STATES

ALASKA

Prince Rupert

Queen Charlotte Islands

Hecate Strait

Queen Charlotte Sound

Northeast Pacific Basin

VANCOUVER

VANCOUVER ISLAND

Juan de Fuca Islands

VICTORIA

SEATTLE

Cape Flattery

Long Island Peninsula

PORTLAND

52°F (11°C)

TSUNAMI HAZARD ZONE

Coos Bay

◇ Eureka

SAN FRANCISCO

Bodega Bay

Monterey Bay

Santa Barbara

◇ Santa Barbara Channel

Pacific Ocean

68°F (20°C)

Northeast Pacific Ocean

 When the navigator *Juan de Fuca* explored the north Pacific coast 430 years ago, it had been inhabited for at least 16,000 years, since the dawn of human habitation in the Americas.

 The *Cadborosaurus*, a mythical and mysterious animal that was thought to live in the waters of British Columbia, was said to have the head of a horse and the body of a serpent.

 The *red sea urchin* can reach almost 8 inches (20 cm) in diameter and can live up to 200 years. Red sea urchins are a favorite food of sea otters.

 The *American shad*, a delicious fish from the American North Atlantic, was introduced into the Sacramento River in 1871 and it has now spread throughout the entire American Pacific.

 The *common dolphin* lives in large schools and moves very swiftly, at speeds up to 24 miles (40 km) per hour.

 The elegant curves of the 4-mile (6.5-km) long *Astoria-Megler Bridge* connect the shores of the mighty Columbia River shortly before it rushes into the ocean.

 The Haida people were famous for their magnificent decorated *canoes* that were made from majestic trunks of red cedar trees and then sold as luxury goods to the heads of other native nations.

 In 1906, a storm tossed the sailing ship *Peter Iredale* onto the coast of Oregon so violently that three of its masts broke. There were no victims, and today the shipwreck is a tourist attraction.

 The *grunion* is extremely widespread in California, where it is caught by fishermen using only their bare hands.

 The *gray whale* travels from Alaska, where it lives during the summer, to Baja California, where it migrates to give birth.

 The *Pacific halibut* is a large flatfish similar to the sole, but with gigantic dimensions. It is more than 8 feet (2.5 m) long and weighs up to 550 pounds (250 kg).

 Beautiful *sea stacks* that look like giant columns of rock can be seen all along the coast of Oregon. They were created millions of years ago by lava flows and the strong waves of the Pacific Ocean.

 According to the mythology of Pacific Northwest peoples like the Kwakwaka'wakw and Nuxalk indigenous nations, *Komokwa* is the lord of the kingdom hidden beneath the ocean. He is also thought to be the master of seals.

 The ocean along the Pacific coast of the United States is so cold that surfers need to wear *wetsuits* to keep warm.

 Two different types of *orca* live around Vancouver Island, resident and transient. The resident populations stay in the area and do not migrate, while the transients only visit the island when they are migrating to other parts of the Pacific.

 The *black brant* is a small goose with a black head and a white collar around its neck.

 The *common seal* is easy to spot on the beaches and rocky coasts of the Pacific, where it enjoys spending long periods lying in the sun.

 The indigenous coastal populations used reeds, plant fibers, and wood to make *baskets* that they used as mini-nets to pull salmon, anchovies, and crabs out of the water.

 The *fishing rod* used to stun salmon was one of the numerous tools used by the ancient inhabitants of the coast.

 Whale watching is a very popular activity along the entire coast.

 Eelgrass in the Puget Sound, the second largest estuary in the United States, offers food, oxygen, and shelter for the underwater life.

The **blue shark** has a recognizable pointed snout and long pectoral fins.

 The **silver salmon** is silver only when found in the ocean. When it begins to swim up rivers to reproduce, its sides become red and its back becomes green.

 In the nineteenth century, **mail boats** running between San Francisco and Hong Kong transported groups of merchants, officials, or American missionaries on the outward journey, and thousands of Chinese immigrants on the return voyage.

 The ancient inhabitants of the California coast used the glossy **mother-of-pearl** from abalone shells to create precious objects like earrings.

 The magnificent **Golden Gate Bridge** in San Francisco is on the list of the Seven Wonders of the Modern World.

 The **sunflower sea star** is one of the largest sea stars on the American Pacific coast. Its color varies from red to pink to yellow and it can have up to 24 arms.

 The **king of herrings** or **giant oarfish** has a flat, ribbonlike shape with red rays on its head, and can grow up to 36 feet (11 m) long.

 The **giant octopus** can be almost 16 feet (5 m) long and can weigh 110 pounds (50 kg). It never attacks humans and actually is afraid of them.

 Amateur fishermen catch about one-third of the **lobsters** that are hunted on the Californian coast.

 The **Cape Mendocino Light** was built in 1868. It was moved to Shelter Cove, California, in 1998.

 The **Farallon Islands**, off the coast of San Francisco, are home to an important wildlife refuge.

 The **Pacific sardine** was one of the first fish species to be threatened by overfishing. It was fished intensively until the 1940s, and its numbers have still not recovered.

 The **pelican** is easy to spot in coastal ports, where it always hopes that fishermen will give it fish, which quickly vanish into its large throat pouch.

 The **Bixby Bridge** is considered the gateway to the spectacular coast of Big Sur.

 The Great Pacific garbage patch is a large patch of trash caught in the northern Pacific current between Hawaii and California. Researchers are looking for ways to reduce the waste, such as **maritime drones** that would collect floating plastic waste.

 For about 200 years, the only contact between Europeans and the coastal populations of Upper California took place in the cape of Point Reyes, where in 1595 the sailing ship **San Agustín** sank along the route between Manila and Acapulco.

 The **white sturgeon** is valued for its caviar and meat. It reproduces in rivers and then returns to the sea to grow.

 The **horn shark** is a small shark with a large head that has two ridges that look like antlers. Its teeth are perfect for crunching sea urchins, mollusks, and crustaceans.

 The **white shark** has always been feared but it is actually in danger of becoming extinct. It is estimated that only a few hundred of them remain in the north Pacific.

 Morro Rock is a small island off the coast of Southern California that was formed by a now-vanished volcano.

 Nine thousand **balloon bombs** were launched from Japan to the United States during World War II.

 Alcatraz was a prison on an island off the coast of San Francisco. It is now a museum very popular with tourists.

 There are a number of **oil rigs** that extend into the Santa Barbara Channel.

 The **spotted ratfish** belongs to the genus *Hydrolagus*, a Latin name that, curiously, means "water rabbit." Despite its shape, it is related to sharks and, like them, has a cartilaginous skeleton.

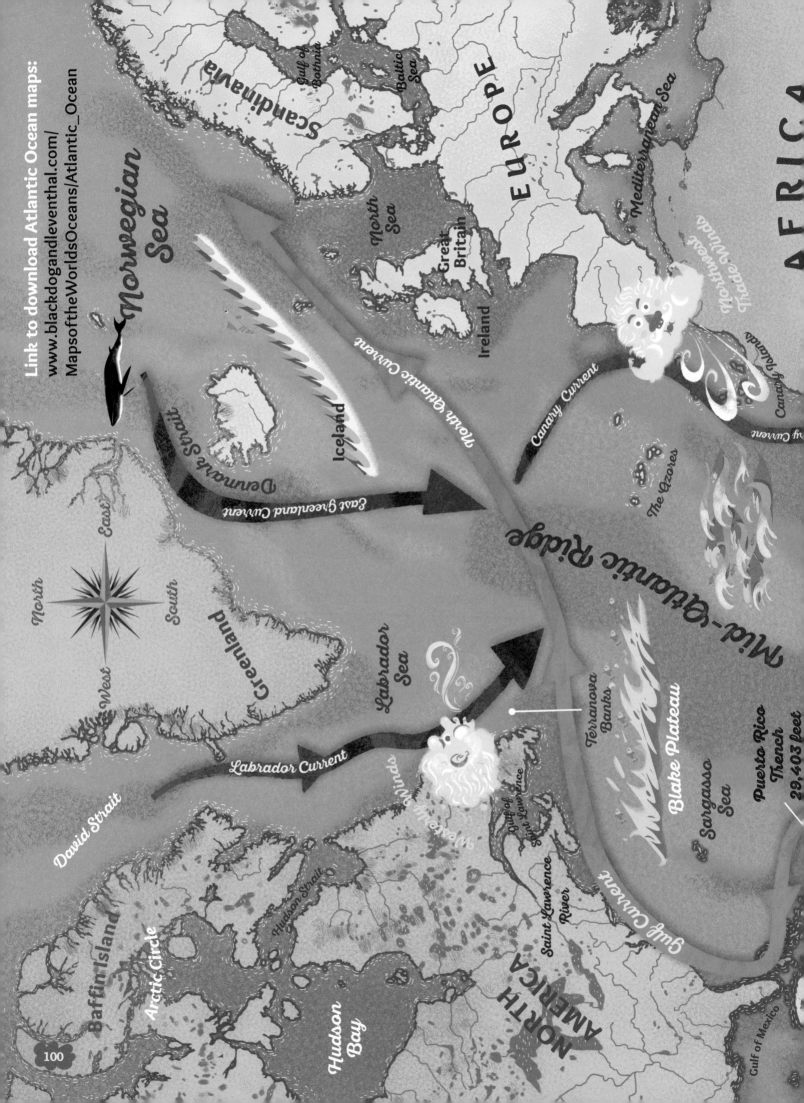

100

Scandinavia

Gulf of Bothnia

Baltic Sea

EUROPE

Mediterranean Sea

AFRICA

Norwegian Sea

North Sea

Great Britain

Ireland

North Atlantic Current

Canary Current

Denmark Strait

Iceland

East Greenland Current

The Azores

Mid-Atlantic Ridge

North

East

West

South

Greenland

Davis Strait

Labrador Sea

Labrador Current

Winds

Terranova Banks

Blake Plateau

Sargasso Sea

Puerto Rico Trench
29,403 feet

Baffin Island

Arctic Circle

Hudson Strait

Hudson Bay

North Winds

Gulf of Saint Lawrence

Saint Lawrence River

Gulf Current

NORTH AMERICA

Gulf of Mexico

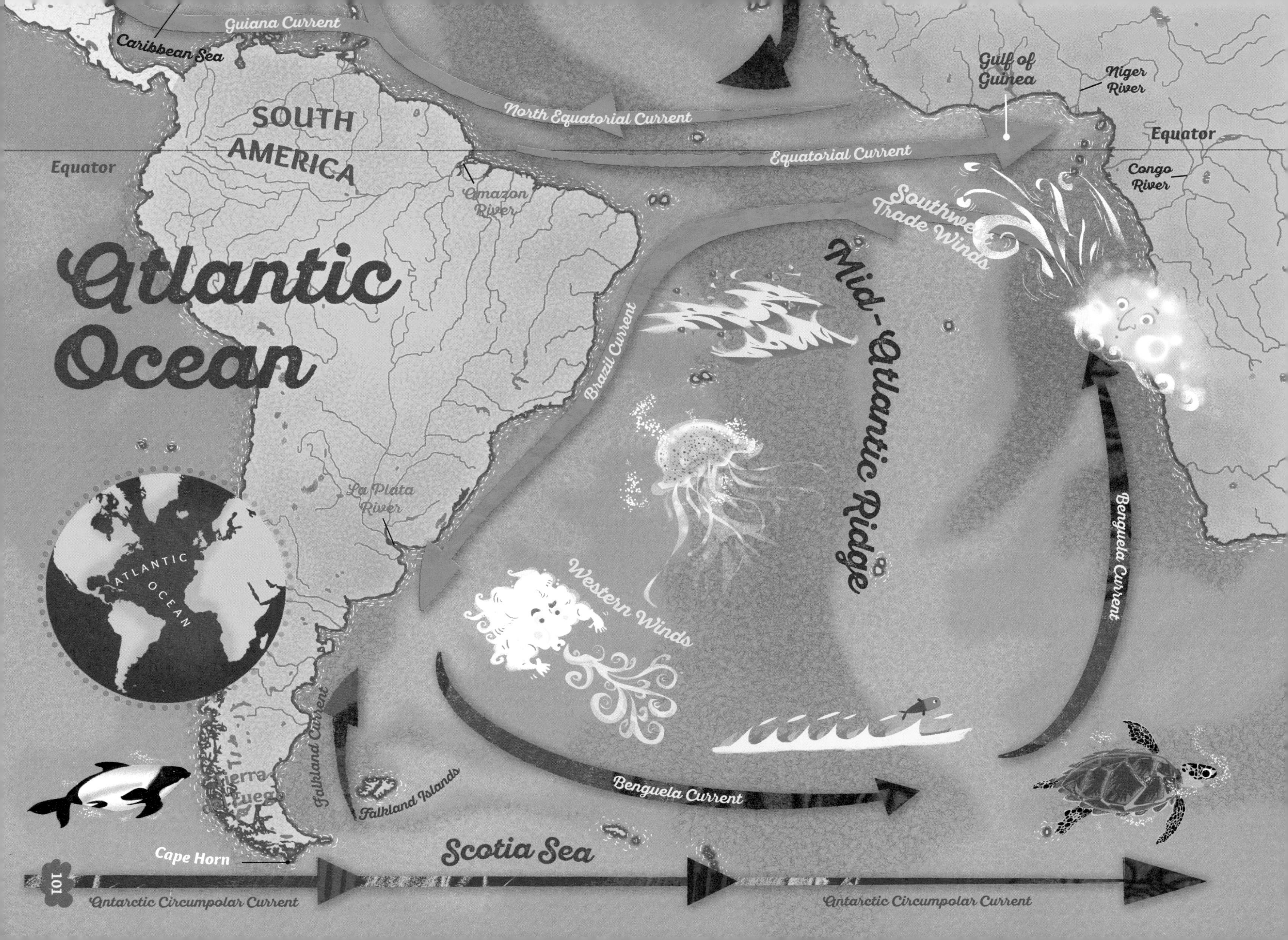

Atlantic Ocean

Benguela Current

Southwest Trade Winds

Mid-Atlantic Ridge

Western Winds

Benguela Current

Scotia Sea

Equatorial Current

North Equatorial Current

Brazil Current

Antarctic Circumpolar Current

Antarctic Circumpolar Current

Equator

Equator

Niger River

Congo River

Gulf of Guinea

Guiana Current

Caribbean Sea

SOUTH AMERICA

Amazon River

La Plata River

Falkland Islands

Falkland Current

Cape Horn

ATLANTIC OCEAN

Atlantic Ocean
Beyond the Pillars of Hercules

The Atlantic Ocean is a vast expanse of deserted waters. It does not have thousands of islands that help with navigation like the Pacific Ocean. This is why it remained unknown and unexplored for a very long time, almost a symbol of the end of the Earth.

Three thousand years ago, the Guanche people discovered the Canary Islands off the coast of Africa. Two hundred years later, the Vikings crossed the ocean, reaching Iceland, Greenland, and North America.

Populations of the eastern coast of the Americas also emigrated to the continent by crossing the Bering Strait, perhaps 20,000 years ago, or even, some say, sailing south of the ice, directly from Europe.

All these peoples were sailors, and in prehistoric times some of them developed vessels and fishing instruments that are still in use today, such as the kayaks and harpoons of the Inuit, or the coracles of the British isles, tiny paddle boats that resemble walnuts.

A great deal of time had to pass before the Atlantic was recognized as an ocean with an elongated shape that separates two large continental masses. Approximately 500 years ago, European cartographers and seafarers grasped its existence, although they thought that the continent across the ocean was Asia.

It is only since 1492, year of the "rediscovery" of America, that the Atlantic would become the setting for undertakings such as circumnavigations of the globe by explorers like Ferdinand Magellan and Francis Drake, in the sixteenth century.

The countless riches discovered in the Americas soon transformed the Atlantic into an enormous commercial highway, across which gold, fruit, and exotic plants traveled, as well as, sadly, millions of slaves, captured in Africa and transported to the other side of the ocean.

These riches also caused wars and piracy. Up until less than 200 years ago, the western Atlantic was a stage for battles between ships sailing for the nations of Europe, and dangerous pirates like Blackbeard.

Today the Atlantic Ocean has lost importance as a commercial route, but it continues to yield riches such as oil, natural gas, and fish. It is the setting for many ancient myths and legends, and is constantly being studied by scientists using research ships.

The Atlantic is the second-largest ocean in the world with a surface area of about 41,100,000 square miles (106,448,511 km^2). The deepest point is the Milwaukee Deep, near Puerto Rico, which is 27,841 feet (8,486 m) below the surface. The Atlantic began forming approximately 150 million years ago and is still expanding at a rate of several centimeters a year.

A large underwater mountain chain, called the Mid-Atlantic Ridge, rises up almost at the center of the Atlantic Ocean. It extends from Iceland in the north to the boundaries of the South Atlantic Ocean in the south. The peaks of the ridge can be more than 8,200 feet (2,500 m) below the surface, but they sometimes emerge, forming islands such as the Azores.

At the height of the Greenland-Iceland Rise, the cold and dense waters of the Arctic sink down, moving southward as far as Antarctica, then branching off into the Pacific Ocean and the Indian Ocean and turning toward the North Atlantic. This path is known as the "global convection" and is important for climate control.

The Sargasso Sea lies at the center of the North Atlantic. Eels have always reproduced here, in an area surrounded by currents. The sea gets its name from the masses of Sargassum seaweed floating on its surface, in which early explorers feared they would remain forever imprisoned.

The warmest and most tropical portion of the Atlantic Ocean is the Caribbean Sea. Located between Florida, Central America, and South America, it is a coral sea with reefs full of sea life, including butterflyfish, parrotfish, whale sharks, rays, and hundreds of other species.

The Atlantic is crossed by significant currents that move millions of cubic feet of warm and cold waters. One of the most important of these is the warm Gulf Stream, which starts in the Caribbean and then moves north and then east, until it reaches and warms the coasts of Europe.

The Atlantic Ocean has always been an important region for fishing. Americans and Europeans fish intensively for cod, tuna, sardines, and mackerel, and for a long time they also fished large cetaceans like whales, blue whales, and sperm whales, as well as various seal species.

The northern portion of the Atlantic Ocean is connected to a great number of adjacent seas, such as the Mediterranean Sea, to which it is linked by the Strait of Gibraltar, the North Sea, the Baltic Sea, and the Irish Sea. The region is home to islands like Iceland, Newfoundland, the British Isles, the Azores, the Canary Islands, the islands of Bermuda, the Antilles, and many others.

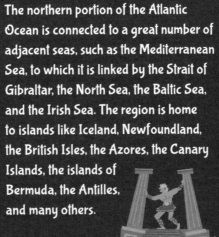

The Atlantic is one of the oceans where the largest tides have been recorded. In the Bay of Fundy and in Frobisher Bay, both in Canada, tides can exceed 70 feet (21 m) and 40 feet (12.3 m) high, respectively. To the south, in Puerto Gallegos in Argentina, high tide can reach 59 feet (18 m).

FLORIDA

Grand Bahama Island

Abaco Islands

◇ Tampa

Miami ◇

Gulf of Mexico

Everglades National Park

NASSAU

Eleuthera

Florida Keys

Bahamas

Cat Isla

Rum C

Exuma

Andros

Long Island

L'AVANA

Sabana-Camagüey Archipelago

Yucatan Strait

CUBA

Isla de la Juventud

Greater Antilles

MEXICO

Mérida ◇

Yucatan Peninsula

Cozumel

Turquino National Park

Yucatan Basin

Santiago de Cuba

Sian Ka'an Biosphere Reserve

Cayman Islands

Cayman Trench

Montego Bay

El Mirador National Park

JAMAICA

KINGSTON

BELIZE

Gulf of Honduras

GUATEMALA

Colombia

HONDURAS

⭑ GUATEMALA CITY

TEGUCIGALPA ⭑

NICARAGUA

Pacific Ocean

Caribbean Sea

MANAGUA ◇

Lake Cocibolca

Barranquilla

COSTA RICA

PANAMA

Panama Canal

32° C

SAN JOSÉ

86°F (30°C)

104

North America

ATLANTIC OCEAN

Central America

CARIBBEAN SEA

PACIFIC OCEAN

South America

Atlantic Ocean

Caribbean Sea

Salvador

ooked and

cklins land

a Islands

Turks and Caicos

Greater Antilles

HAITI

DOMINICAN REPUBLIC

PORT-AU-PRINCE

SANTO DOMINGO

Puerto Rico Trench

Milwaukee Deep

SAN JUAN

PUERTO RICO

British Virgin Islands

U.S. Virgin Islands

Anguilla

St. Kitts and Nevis

Montserrat

Antigua and Barbuda

Guadeloupe

Dominica

Martinique

St. Lucia

Basin

Venezuela Basin

Lesser Antilles

Saint Vincent and the Grenadines

Barbados

Aruba

Bonaire

Curaçao

Blanquilla Island

Grenada

Gulf of Venezuela

Los Roches Archipelago

Tortuga

Margarita Island

PORT OF SPAIN

Tobago

COLOMBIA

Maracaibo

Valencia

CARACAS

Trinidad

Lake Maracaibo

V E N E Z U E L A

Caribbean Sea

 In 2010, human error caused an **oil spill** in the Gulf of Mexico that lasted almost 5 months, the most serious marine oil spill in history.

 Some have claimed to see the **lusca**, a many-headed sea monster with tentacles dozens of feet long, in the Bahamas, but they may have just seen a giant octopus.

 You don't need a time machine to see the **pirate ships** of the Caribbean. Today, modern galleons transport tourists on adventure cruises.

 Fishing with a **trident harpoon** or other shapes of harpoons is an old fishing technique, used both in Caribbean waters and in the large rivers of the Amazon region in South America.

 A **research ship** is studying the submerged Chicxulub crater, where, about 66 million years ago, a comet or asteroid crashed, perhaps causing the end of the dinosaurs.

 The **Lighthouse Reef** in Belize is considered one of the best dive sites in the Caribbean. It is close to the Great Blue Hole, which is almost 490 feet (150 m) deep.

 When it is time to reproduce, the **Nassau grouper**, a symbolic species of the Caribbean, can swim more than 124 miles (200 km) to gather in large schools.

 In 1962, the Caribbean Sea was full of military submarines, warships, and nuclear missiles during the **Cuban Missile Crisis**, which brought the world to the brink of world war.

 Stingray City is located in the shallow waters of the Grand Cayman Islands. It is known throughout the world because people can swim here among large rays that seem to fly underwater.

 GlobeNet is a 14,600-mile (23,500-km) long cable that transports computer data between continents, passing along the bottom of the oceans.

 The **Zapata Swamp** in southern Cuba is a biosphere reserve. Rare species such as the Cuban crocodile, the alligator gar, and the bee hummingbird find shelter here.

 The **great barracuda** grows up to 5 feet (1.5 m) long and is a sea predator. Its large mouth is equipped with pointed, sharp teeth that can tear a large fish in half.

 The **blue marlin** prefers to swim in the open ocean, far from the coasts. It can move at speeds up to 68 miles per hour (110 kph) and can cover up to 43 miles (70 km) in a day.

 At sunrise or sunset, the sun gives off an emerald-green glare for a few moments, known as the **Green flash** phenomenon.

 A statue of the mysterious **Chacmool**, an ancient mythological figure connected to sacrifices, was found in the waters off the Mexican island of Cozumel.

 The **Putún** people, the Phoenicians of pre-Columbian America, built long boats they used for trading in tobacco, salt, gold, and cocoa in the Caribbean islands.

 Peter Martyr d'Anghiera was a historian who never crossed the ocean, but in 1493, he was the first person to become convinced that the Caribbean had already been discovered before the time of Columbus.

 The wealth of food in the Caribbean Sea is exploited by fleets of **small fishing boats** that usually venture less than 12.5 miles (20 km) from the coast.

 Legend has it that a **whirlpool** indicates the point where the beautiful Anuanaitu drowned and transformed herself into the Soul of the Ocean.

 The **giant Caribbean sea anemone** attaches itself to the ocean floor, revealing only its crown of colorful tentacles, which can reach a diameter of almost 20 inches (50 cm).

 The **Caribbean spiny lobster** can grow 2 feet (60 cm) long and it is protected by a coffee-colored exterior with yellow and black patches. It also has two sharp spines to protect its eyes.

 The **scorpionfish**, a species typically found in the Indo-Pacific waters, was transferred by some aquariums into the waters of the Caribbean, where it is becoming invasive and a threat to local fish.

 The **Caribbean reef shark** is the most common shark in the Caribbean and can be seen within the first 100 feet (30 m) underwater, while it swims on the reef.

 Mangroves, which make up one of the most significant coastal ecosystems in the Caribbean, are found from Florida to Cuba, from the Bahamas to Jamaica, and south to Venezuela.

Elkhorn coral grows very rapidly, and is one of the most important organisms that build the Caribbean reefs, particularly those closer to the surface.

 Successful films and books recount terrible stories of **shark attacks** in the Caribbean Sea. However, in places like Bermuda, there have only been three shark attacks in the past 60 years.

 At 155 miles (250 km) long, the **Belize Barrier Reef** is the largest reef in the Caribbean, and is the second largest in the world, after the Australian Great Barrier Reef.

 Whale sharks are frequently seen in the Caribbean Sea, where, in the summer months, they meet along the coasts that are full of Yucatan plankton, before heading out to the open sea.

 The **Caribbean manatee** is an aquatic mammal that feeds on underwater plants. It has intestines that can be up to 148 feet (45 m) long.

 According to legends, wonderful **riches** were transported and lost centuries ago in the Caribbean, and tons of gold still lies hidden somewhere.

 Some fifteenth-century maps show **Antillia**, a phantom island that was never discovered, to the west of Portugal and Spain.

 The **Caribbean flying fish** has long pectoral fins that look like wings. They can leap 4 feet (1.2 m) above the water and glide in the air for 655 feet (200 m) before returning to the sea.

 Falcons, falconets, and carronades are types of naval cannons that are still seen on the ancient fortifications of the Caribbean. While very imprecise, they were dangerous and devastating.

 The pirate Jack Sparrow from the **Pirates of the Caribbean** movies is imaginary, but may have been inspired by real-life pirates like Henry Morgan or Blackbeard.

 The ancient Mayas living along the coast were fishermen, as well as farmers. Their fish-god, **Chac Uayab Xoc**, protected those who went out to sea to gather food.

 The **Miskito** people of Nicaragua are skilled sailors who use canoes suitable to the open ocean. They formed an independent kingdom until 1894, with their own kings, governors, and admirals.

 During storms, ships are sometimes enveloped by ghostly lights known as **St. Elmo's fire**, a natural phenomenon that ancient sailors considered good luck.

 The Brazilian **jangada** is a sailboat that has prehistoric origins, but is still very useful today. Craftsmen make these boats using only wood and fiber—not even one nail.

 The **blackfin tuna**, or Atlantic tuna, is a small species of tuna that, together with skipjack tunas, form schools that can include thousands of fish.

 The **aycayía**, mermaids of the Caribbean, are thought to be sometimes good and sometimes evil. Their name means "girls with beautiful voices."

 The **humpback whale**, with large wings along its pectoral fins, can emerge almost completely from the water.

 The **queen conch** is a large-shelled mollusk that is fished by the thousands every year for its meat.

 The **bull shark** is one of the most dangerous sharks in the world. They can grow up to 11 feet (3.5 m) and can weigh 500 pounds (227 kg).

 In Costa Rica, programs to **rescue turtles** are run by volunteers who protect the animals and offer a source of income for local inhabitants.

 The name **Caribbean** comes from the **Caribs**, another ancient population of sailors who spread out from the Amazon on board large canoes, invading and conquering the Antilles Islands.

 The **American crocodile** can reach up to 20 feet (6 m) long and can weigh up to 1 ton (0.9 metric ton). A fearsome predator, it has almost no enemies when it reaches adulthood.

Caribbean:
A Sunken Paradise

The coral reefs of the Caribbean Sea are some of the most beautiful and distinctive marine areas of the Atlantic Ocean. The reefs are home to dozens of different corals and hundreds of species of fish. The coral reefs also protect coastlines from the damaging effects of wave action and tropical storms.

Northwest Atlantic Ocean

Anticosti Island

Honguedo Strait

Saint Lawrence Gulf

Magdalen Islands

Cape Breton Island

Cape Canso

46°F (8°C)

Prince Edward Island

HALIFAX

Nova Scotia

Northumberland Strait

Gaspé Peninsula

Chaleur Bay

Saint Lawrence River

Grand Manan Island

Bay of Fundy

Cape Sable

CANADA

Québec

Gulf of Maine

Cape Cod National Park

Portland

Boston

JAWS

WOODS HOLE OCEANOGRAPHIC INSTITUTION 1930

Hudson River

UNITED STATES

New York

Long Island

Hudson River

Philadelphia

Atlantic City

Delaware Bay

Baltimore

WASHINGTON

Susquehanna River

Lake Phelps

Chesapeake Bay

Chowan River

ATLANTIC OCEAN

Canada

North America

Central America

SARGASSO SEA

South America

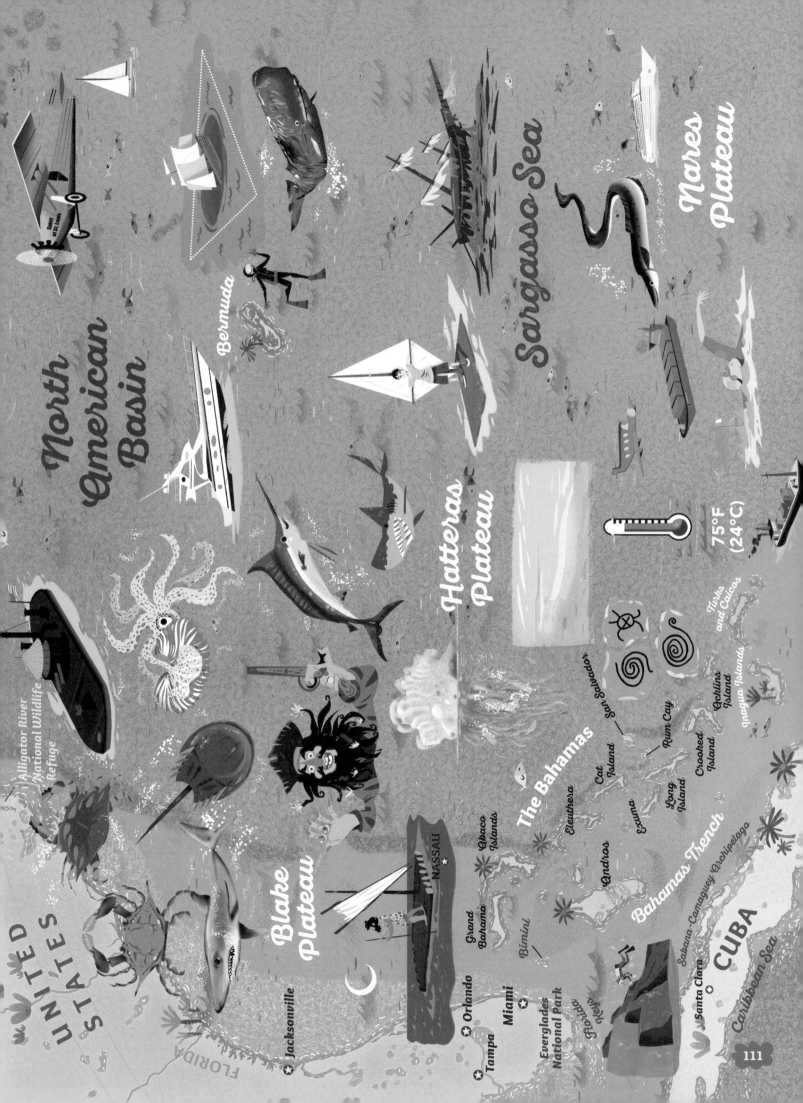

North
American
Basin

Sargasso Sea

Nares
Plateau

Bermuda

Spirit
of St. Louis

75°F
(24°C)

Hatteras
Plateau

Alligator River
National Wildlife
Refuge

The Bahamas

Turks
and Caicos

San Salvador

Rum Cay

Acklins
Island

Inagua Islands

Cat
Island

Crooked
Island

Blake
Plateau

NASSAU

Eleuthera

Long
Island

Exuma

Bahamas Trench

Abaco
Islands

Andros

Sabana-Camaguey Archipelago

Grand
Bahama

Bimini

CUBA

UNITED
STATES

Orlando

Tampa

Miami

Everglades
National Park

Florida
Keys

Jacksonville

Santa Clara

Caribbean Sea

FLORIDA

111

Northwest Atlantic Ocean

The *eastern oyster* is a bivalve mollusk with a large, flat shell. It is a typical species of the American Atlantic coasts and can reach up to 8 inches (20 cm) long.

The inspiration for the famous shark in the 1975 movie *Jaws*, which took place off the coast of New England, may have been a bull shark.

The *harbor seal* is one of the most widespread and well-known seals thanks to its chubby appearance, doglike nose, and large black eyes.

Since the Europeans discovered America, this region has been known for its abundance of fish and *fishermen*.

The *American lobster* is the heaviest crustacean in the world, and can grow to more than 2 feet (60 cm) long and can weigh 44 pounds (20 kg). At birth, they weigh only a tenth of a gram.

The *lion's mane jellyfish*, also known as the giant jellyfish, is the largest known jellyfish in the world. It stings but is not fatal to humans.

The *Minas Basin* is located at the entrance to the Bay of Fundy in Nova Scotia in Canada. It is a vast, flat, sandy area visited by hundreds of thousands of waterfowl every year.

People have searched for hidden treasure in the *Oak Island pit* in Nova Scotia for hundreds of years, but no one has been able to reach the bottom because seawater floods their attempted excavations.

The *European hake* is shimmering gray and silver in coloring. It is an extremely voracious predator and feeds predominantly on herring.

The *Old Sow* whirlpool forms in Passamaquoddy Bay in Canada, where the ocean currents flow among the islands at 30 miles per hour (48 kph), creating underwater twisters up to 100 feet (30 m) in diameter.

The *Kemp's ridley sea turtle* is the rarest and smallest sea turtle in the world. It is also the only turtle that nests during the day.

The *menhaden* is a typical coastal fish that also swims upriver when it is young. It mostly feeds off of plankton that it filters from the water through its spiny gills.

The *common sun star* looks like a small sun thanks to its many arms (up to 14).

Viking ships that landed in America 500 years before Columbus were extraordinarily practical. They had prows and sterns of equal length that could change direction almost instantly.

About 35 million years ago, a meteorite fractured the Earth's crust, reaching a depth of 5 miles (8 km), creating a large *impact crater* in the Chesapeake Bay, near the states of Maryland and Virginia.

The *osprey* is a large bird of prey that specializes in feeding on fish. It has sturdy wings and rough-toed feet with long, curved claws that it uses to hold on to slippery prey like fish.

The *longfin inshore squid* is a speedy predator, extremely skilled at capturing crustaceans and fish with its extendable and elastic tentacles.

Lobster traps are wooden cages covered in fishnets that were invented more than 200 years ago in Massachusetts. They are still used by amateur fishermen today.

The *Woods Hole Oceanographic Institution* is a famous research center, founded in 1930, that studies oceans in all fields of marine sciences, from marine biology to oceanography.

The *Titanic* sank 375 miles (600 km) south of Newfoundland after hitting an iceberg in 1912. It came to rest about 12,500 feet (3,810 m) below the surface.

In the sixteenth century, explorers and sailors like the Englishman *Sir Walter Raleigh* were sent across the ocean by their kings and countries to establish colonies.

The **great white shark**, the most famous of all sharks, is quite common along the American coasts, where it has been declared an endangered species because of its importance to ocean life.

 There are numerous shipyards along the western coast of the Atlantic that build and repair luxury **yachts** that sail throughout the world's oceans.

 More than 50 years ago, S. Newman Darby invented the **sailboard** to windsurf the small waves on the lake behind his house in Pennsylvania. It is now used all around the world.

 The **sperm whale** is the largest cetacean with teeth, and the mammal that can swim to the greatest depths. One whale was detected at a depth of 7,380 feet (2,250 m).

 The **Bermuda Triangle** is an unsolved ocean mystery, an area where many ships and airplanes are said to have disappeared.

 The **Portuguese man o' war** moves with the currents like a jellyfish, but it floats on the surface thanks to powerfully stinging tentacles that hang down from a gas-filled, violet-colored sac.

 The **green crab** eats the widest variety of marine plants and animals compared to other crabs. Green crabs can also live in a variety of environments.

 The **mako shark** is capable of traveling more than 1,675 miles (2,700 km) in the open ocean, and can swim at speeds of more than 46 miles per hour (74 kph).

 Pink Sand Beach in the Bahamas is pink thanks to the collection of colorful shells of tiny organisms called foraminifera.

 In 1973, the wreck of the **USS Monitor** was discovered off Cape Hatteras, North Carolina. It was used in the first combat between two armored battleships, during the American Civil War, and sank in 1862 in a storm.

 Some people believe that **prehistoric populations** left Europe approximately 20,000 years ago and arrived in America, traveling along the edge of the ice that covered the northern Atlantic.

 A mystery of the Bahamas is the **Bimini Road**, an underwater stone structure discovered in 1968, believed by some to be human-made and very old (3,500 to 15,000 years old).

 The first human being to cross the entire Atlantic by airplane was Charles Lindbergh, who carried out his historic flight in 1927 on board the **Spirit of St. Louis** monoplane.

 The waters around Florida were dominated by the fearsome pirate **Blackbeard**, who was killed in 1718.

 Water, turtles, and shells are three of the symbols of the **Taíno** people, the first islanders Christopher Columbus encountered in the Americas. They used these symbols to tell their stories by carving them into stone.

 The **blue crab** has a distinctive large shell, bluish color, and back claws shaped like shovels, which help it swim.

 The **white marlin** has a sword like a swordfish. It can grow up to 9 feet (2.7 m) long and usually stuns its prey with blows from its tail before eating it.

 Frenchman **Benoît Lecomte** claims to have swum across the Atlantic in 1998, covering about 3,700 miles (5,955 km) in 73 days.

 The **limulus**, also called a horseshoe crab because of the shape of its shell, is a large, compact crustacean with a long, rigid tail that it uses as a lever to get back on its feet after being knocked over.

 The **Sargasso Sea** is the only sea in the world without land boundaries.

 The **argonaut** is an octopus whose males and females come in very different shapes. Males, which are very small, have an octopus shape, while females live inside extremely beautiful shells.

 Eels are born in the sea and then grow in rivers before migrating back for thousands of miles to reach the depths of the Sargasso Sea.

Southern Atlantic Ocean

81°F (27°C)

Santos Plateau

Atlantic Ocean

Argentine Basin

Salvador

BRAZIL

⭐ BRASILIA

Rio De Janeiro

BOLIVIA

PARAGUAY

Porto Alegre

Patos Lagoon

URUGUAY

MONTEVIDEO ⭐

Rio de la Plata Delta

Rio de la Plata

BUENOS AIRES ⭐

ARGENTINA

SANTIAGO ⭐ DE CHILE

CHILE

Pacific Ocean

ATLANTIC OCEAN

SCOTIA SEA

South America

PACIFIC OCEAN

CHILE

San Matías Gulf

Patagonia Platform

PATAGONIA

San Jorge Gulf

Baía Grande

Strait of Magellan

Punta Arenas

Tierra del Fuego

Ushuaia

Isla de los Estados

Wollaston Islands

Cape Horn

Drake Strait

34°F (1°C)

South Shetland Islands

South Orkney Islands

Scotia Sea

South Sandwich Islands

South Georgia Island

Falkland Plateau

Falkland Islands

Sailing Alone Around the World

Captain Joshua Slocum

Southern Atlantic Ocean

 Between the sixteenth and twentieth centuries, millions of **African slaves** were transported across the Atlantic, chained together on board merchant vessels.

 Cargo ships similar to **floating trucks** stock dozens of drilling facilities along the Brazilian coasts, where 12 oil fields are found.

 The **Argentine anchovy** lives in dense schools, between 98 and 295 feet (30 and 90 m) below the surface. When they reproduce, they can be found more than 500 miles (800 km) from the coast.

 The large ocean waves that bathe the coast of Brazil are perfect for **surfing**, but also for **bodysurfing**, where people ride the waves without a board.

 Rusted fishing boats usually are not shipwrecks. They were abandoned by their owners for financial reasons.

 Brilliantly colored fishing boats have a peaceful air, but they are often involved in **fishing wars** that flare up among fleets from different countries.

 Sixteenth-century Portuguese historian Pero de Magalhães Gândavo documented the fauna of Brazil and also mentioned **Ipupiara**, a human-eating sea monster feared by the native Tupi population.

 Sardines were once abundant along the South American coast, but are now decreasing in number.

 South American fur seals generally do not like beaches. They prefer rocky cliffs, which they can climb thanks to their strong fins.

 According to a Brazilian legend, the **Great Sea Serpent** was said to have created night with the darkness of the ocean floor, when his daughter went to live on land.

 By attaching **transmitters** to the shells of newborn turtles, biologists can study their movements across the Atlantic, and are making interesting discoveries.

 Hairy **lures**, similar to flies, are used for fishing trout and salmon along the coasts and in the estuaries of Argentine rivers, where these fish are plentiful.

The **sharpnose sevengill shark** has seven gill slits instead of five, like other sharks. It usually lives in deep, dark waters, which is why it has large eyes.

The **La Plata dolphin** lives in the coastal Atlantic waters of South America. It usually lives in murky waters and uses sound waves to find its prey.

The **Falkland sprat** lives in large schools and is considered the most plentiful fish species in this part of the Atlantic.

 Yemoja is the good queen of the sea and the protector of fishermen and castaways, according to the Candomblé religion, which was practiced in Brazil by African slaves.

 Male **South American sea lions** can weigh almost twice as much as the females, 770 pounds (350 kg) versus 330 pounds (150 kg), and they have light manes that cover their heads.

 The most ancient history of the oceans can be studied thanks to **logging instruments**, large drills that are lowered to extract samples of sediments deposited on the floor of the ocean over millions of years.

 The **crabeater seal** does not actually eat crabs. It eats krill, which make up 90 percent of its diet.

 Although it is thousands of miles away from Europe, in 1939, Río de la Plata was the scene of the first **naval battle** in World War II.

 A type of **fish trap** made from flexible wood is found throughout the world. Once fish enter, they have no way to get out.

The oceanographic ship **RV Atlantis II**, no longer in service today, carried out important research throughout the world, traveling nearly 1,242,000 miles (2 million km)!

The Yámana people used **bark canoes** resembling walnut shells to hunt seals, whose fat was essential for surviving in frigid Tierra del Fuego in Argentina.

The strait that separates South America from the Antarctic peninsula and joins the Atlantic and Pacific oceans, the Drake Passage, is named for **Sir Francis Drake**, who crossed this body of water in 1578.

The **southern right whale** has an enormous tail, which it can hold stretched out of the water to be pushed by strong gusts of wind, known as tail sailing.

The **crested penguin** has a yellow-and-black plumed crest. For much of the year, it lives in the sea, hunting krill and other crustaceans.

The **Shag Rocks** are six small, uninhabited islands 150 miles (240 km) west of the island of South Georgia that look like shark teeth, but take their name from the birds (a type of cormorant) that are their sole inhabitants.

The **chrysaora lactea** is one of the most common jellyfish along the Atlantic coasts of South America. In the summer, it can form dense, stinging groups that create serious problems for swimmers.

The **blue whale** is not really blue. It is mostly gray, but looks blue because of the reflections of the water.

The **Commerson's dolphin** is easily recognizable thanks to its black head, dorsal fin and tail, and white body.

The **Galathea gregaria** is a crustacean that looks like a small lobster, but is also similar to a hermit crab.

The **wandering albatross** uses its large wings to fly like a glider, taking advantage of air currents for days without ever touching down on the ground, and capturing squid while these birds fly over the surface of the water.

The **painted shrimp** is a small cold-water crustacean that is born male and as it grows larger becomes female.

Magellanic penguins gather every year on the Punta Tombo peninsula in Patagonia, where almost a million pairs meet up to reproduce, always returning to the same nests.

Male **elephant seals** have trunk-like noses, while females, which are smaller, have pointed noses.

Less known than Magellan or Drake, **Joshua Slocum** was an equally extraordinary seafarer who circumnavigated the globe in the late nineteenth century alone in a sailboat.

The **marine otter**, also known as a "marine cat" because of its agility, is about 3 feet (1 m) long and has a double layer of hair, the outermost of which insulates it from the cold water.

The **Patagonian hake** lives in the southern part of the western Atlantic in an area of the ocean that is full of fish. It can be more than 3 feet (1 m) long and its mouth is entirely black.

Krill are small crustaceans similar to small shrimp. They are usually around 3 inches (7 cm) long and can be found in oceans all over the world.

Naval ships used for Antarctic patrols such as the British **HMS Endurance** have red-painted hulls, or bodies, to make them more visible.

Some **icebergs** transported by the current into the South Atlantic can be up to 62 miles (100 km) long, much larger than those found in the North Atlantic.

The **Argentine hake** is a silvery-golden fish with an elongated body, large head, pointed snout, and prominent jaw.

The **southern king crab** is a large red crab that is important for fishing. In the 1970s it nearly caused a war between Argentina and Chile.

The Titanic Wreck

The *Titanic* hit an iceberg off Newfoundland, Canada. It now lies down on the bottom, about 12,500 feet (3,810 m) below sea level.

Despite the tragedy, today the ship is full of life, thanks to the sea creatures that live and hunt inside the sunken ship on the ocean floor.

The sunken ship has been explored by various underwater robots beginning with Dr. Robert Ballard's historic Argo radio-controlled camera in 1985.

The RMS *Titanic* was a British cruise ship that sank on April 15, 1912, during its first voyage from England to New York after hitting an iceberg. More than 1,500 people died.

12,500 feet (3,810 m)

Irish Basin

Faroe Islands

Hebrides Platform

Orkney Islands

Mainland

Hebrides Island

Scotland

GREAT BRITAIN

Glasgow

Edinburgh

Rockall Trench

48°F (9°C)

Isle of Man

Irish Sea

DUBLIN

IRELAND

Liverpool

Celtic Sea

Wales

West European Basin

English

Guer

Jers

Bay of Biscay

PA13-22

Norwegian Sea

Norwegian Trench

Shetland Islands

Bergen

SWEDEN

Gulf of Bothnia

32°F (0°C)

OSLO

NORWAY

STOCKHOLM

Gotland Basin

North Sea

Skagerrak Strait

Gothenburg

Gotland Island

Jylland

DENMARK

Baltic Sea

COPENHAGEN

Helgoland Gulf

Odense

Bay of Kiel

Frisian Islands

Kiel

Hamburg

THE NETHERLANDS

England

AMSTERDAM

BERLIN

LONDON

Strait of Dover

GERMANY

Rhine River

Thames River

Channel

BRUSSELS

BELGIUM

Meuse River

Northeast Atlantic Ocean

PARIS

FRANCE

Seine River

Nantes

Loire River

NORTH SEA

United Kingdom

Norway Sweden

ATLANTIC OCEAN

Europe

Africa

Northeast Atlantic Ocean

 Scapa Flow is a body of water in northern Scotland that is home to an enormous ship cemetery.

 The **harpoon cannon** was invented in the nineteenth century to hunt whales.

 The maritime countries of northern Europe first developed **ferries** that run on **liquefied natural gas**, which is less harmful to the environment than diesel.

 The **Arctic cod** has always been one of the most prized fish. It is consumed fresh, preserved in salt (baccalà), or dried (stockfish).

 The **grampus**, or Risso's dolphin, is a large dolphin with a squared-off head and dark coloring. Older grampuses have many light scars over their skin.

 The **gulper shark** inhabits the eastern Atlantic, where it is most frequently sighted at depths greater than 3,280 feet (1,000 m). Dark in color, it has rough skin and can grow up to 3 feet (1 m) long.

 The **great skua**, a large seabird, is a predator of fish but eats other things as well. They are aggressive birds that sometimes kill smaller birds like puffins.

 The **Atlantic salmon** initially lives in rivers, but between the ages of one and four it swims downstream to the sea, where it remains for another four years before swimming back up the rivers to reproduce.

 Some **Irish and Scottish monks**, fleeing barbarian invasions around 1,500 years ago, transformed themselves into sailors, discovering islands such as the Faroe Islands and Iceland.

 The **Nordland** is a type of fishing boat with an unusual square sail that was used for centuries in northern countries like Norway.

 The **Oseberg ship** is a Viking ship that was discovered almost intact in 1904, an extraordinary example of an actual Viking ship—rapid, maneuverable, and versatile.

 Eelgrass is a perennial ocean plant that can form extensive underwater prairies along the Atlantic coasts, on sandy and muddy ocean floors close to the surface.

 The **white-plumed anemone** resembles coral and mother-of-pearl, but it has a skeleton of soft tissue that can expand, absorbing water and transforming into a small feathery tree.

 Stockfish is air-dried fish, usually cod, that can be preserved for long periods of time.

The **twait shad** is a migratory fish that swims up rivers to reproduce. It can grow up to 24 inches (60 cm) long and is silvery in color with dark patches on its sides.

 The **Arctic tern** is an extraordinary flier and holds the record for the longest migration. It migrates from the northern oceans to Antarctica and then back again every year.

 The **harbor porpoise** is the most frequently sighted cetacean in the North Sea.

 Japanese wireweed is an invasive species that is widespread in the North Sea. It can grow up to 33 feet (10 m) long and can sometimes interfere with boats in bays and ports.

 An unidentified object known as the "**Baltic Sea anomaly**" was discovered at the bottom of the Baltic Sea in 2011. Some people think it's a meteorite or volcanic formation, and others think it's a secret weapon or spaceship.

 Cable ships were developed in the nineteenth century to install the first telegraph lines between Europe and America. They are still used to position modern fiber-optic cables.

 The **herring** is probably the most well-known fish in the North Sea, where it has been caught since ancient times. Herring fishing is now strictly regulated.

 The **zebra mussel** is named for its striped shell and is an invasive bivalve mollusk that has traveled from rivers to the less salty portions of the Baltic, entering into competition with other mollusks.

 The **sea scallop** is one of the most prized bivalves and is both caught and cultivated. Its shell is famous for being the symbol of pilgrims who follow the Camino de Santiago, or Way of St. James.

 The **Pesse canoe** is believed to be the most ancient known boat in the world. It was found in the Netherlands in 1955 and is thought to have been built between 8040 and 7510 BCE. The canoe is almost 10 feet (3 m) long.

Matthew Maury is an American pioneer of oceanography from the nineteenth century whose writings and research about sea currents and meteorology are still being used today.

The **sea lamprey** is considered a primitive fish because it has no jawbone. It lives on other fish, attacking with its suction cup–like mouth that is covered with hornlike teeth.

The lost island of **Hy-Brasil** was shown on various maps until 1865, west of Ireland, but no one has been able to find it.

The **monkfish** mostly lives on the ocean floor where it is very well hidden. Its special dorsal shaft acts as bait for prey.

The **Atlantic wreckfish** can grow up to 7 feet (2 m) long. When they are young, they stay close to the ocean's surface before settling on the ocean floor when they are adults.

The **fucus**, or sea oak, is a brown algae that can be found from the Atlantic to the Baltic. It grows up to 35 inches (90 cm) long, is dark green in color, and covers rocks in dense carpets that are visible at low tide.

The **harbor seal**, struck by a serious virus at the end of the last century, has made a comeback and is once again one of the most common seals in the North Sea, where it reproduces.

A sunken German U-boat was found in 2015 in the northern Atlantic. It may be the sunken **UB-85 submarine** that some believe was attacked by a sea monster in 1918.

The strange-looking **lumpsucker** has a massive body without scales, but is covered with tubercles. It is usually bluish-gray, but during periods of reproduction the males turn orange or red.

Many of the worst naval disasters occur during war, such as the wreck of the **Lusitania**. The transatlantic liner sank near Ireland in 1915, killing 1,198 people.

The **brown crab** is a large crab that can grow up to 10 inches (25 cm) wide.

The **European plaice** is a flatfish. It is an important resource in the North Sea and one of the most frequently caught fish in this region.

Along the flat, sandy southern coast of the North Sea, **shrimp fishermen** fish on horseback along the beach, throwing out their nets.

Many people attempt to swim **across the English Channel**, which is 21 miles (34 km) wide. The first person successfully swam it in 1875 and it took 21 hours, 45 minutes.

Robert FitzRoy, in addition to being the commander of the ship that brought Charles Darwin around the world, invented various types of **barometers** to make sailing safer.

Whale sharks can often be seen along the English coasts, from Cornwall to Scotland.

A **knarr**, an ancient Norse merchant ship, appears on the seal of the Hanseatic League, a powerful trade organization that operated in the Middle Ages between the Baltic and the North Sea.

Fire ships were unmanned ships that were set on fire and sent out against enemy fleets during battle. This was how Spain's famous **invincible Armada** was destroyed in 1588.

Falmouth Bay in Cornwall is home to abundant sea fauna, which finds an ideal environment in the forests of seaweed that grow on the submerged rocks.

The largest fleet of ships ever assembled, 6,939 vessels, were involved in the **Invasion of Normandy** in 1944 during World War II.

The most thrilling flyovers in history have been carried out over the ocean, beginning with **Louis Blériot**'s crossing of the English Channel by airplane in 1909.

Flat, concave **oysters** are fished and cultivated along the French Atlantic coasts, as well as in the North Sea.

Northern Atlantic Ocean
Greenland Sea and Norwegian Sea

Greenland Sea

ATLANTIC OCEAN

Greenland

GREENLAND SEA

Europe

Africa

GREENLAND

Kangerlussuaq Fjord

Ittoqqortoormiit

Cape Brewster

Gunnbjørn Fjeld

Canal

28°F (-2°C)

Iceland Plateau

Arctic Circle

King Frederick VI Coast

Tasiilaq

Denmark

Kg 23
KG 57
Kg 104
Kg 154

ICELAND

REYKJAVÍK

Vatnajökull National Park

Reykjavík Basin

Irminger Basin

Reykjanes Ridge

Cape Farewell

Icelandic Basin

124

Jan Mayen Fracture Zone

Jan Mayen

Jan Mayen Ridge

Norwegian Sea

Norway Basin

Lofoten Islands

Arctic Circle

Faroe Islands

Faroe Shetland Channel

Shetland Islands

Norwegian Trench

Trondheim

Bergen

NORWAY

SCANDINAVIA

SWEDEN

STOCKHOLM

OSLO

Northern Atlantic Ocean

 Adult **harp seals** are not well known, but their pups are famous for their dense, warm white coat, which protects them from the cold.

 The **northern bottlenose whale**, a large cetacean up to 32 feet (10 m) long and weighing as much as 8 tons (7 metric tons), swims northward toward the edge of the Arctic Sea in search of squid.

 The **sjøtrolls** are "evil trolls of the sea" from Nordic folklore. They live far from humans—except for those who have the misfortune to encounter them.

 The **bowhead whale** has an enormous head compared with other whales, and can be up to 59 feet (18 m) long.

 Many of the 3 million **shipwrecks** estimated to exist in the world lie in the Norwegian Sea, most dating back to World War II.

 Stone pilings many feet high, seen in the fjords of Norway, are ingenious traps from which a net is lowered to capture salmon.

 In the sixteenth century, Swedish writer Olaus Magnus had great success in Europe with his astounding descriptions of the northern seas that were populated with **sea monsters**.

 The god **Njorðr**, more peaceful than other ancient Nordic divinities, was the benevolent god of the sea, much loved by sailors and most of the Vikings.

 The people of Iceland used to have to **lift stones** weighing from 50 to 341 pounds (23 to 154 kg) to prove that they were able to withstand the harsh life of being a fisherman.

 Some people claim that the **ghostly hand** photographed in the Greenland Sea in 2013 was a mermaid, but it was just a prank.

 The **northern prawn** is a crustacean that is actively fished in the deep, cold waters of the North Atlantic.

 The **Atlantic puffin** is a seabird that feeds on small fish, which it follows underwater, using the thrust of its strong, stocky wings.

 Icebergs that break away from the ice in Greenland have such spectacular shapes, they are a tourist attraction.

 The **maelstrom** is an ocean current that creates amazing whirlpools. The most powerful one in the world is in the **Saltstraumen**, in Norway.

 Chimaera are cartilaginous fish similar to sharks. Their first dorsal fin is mobile and they have a sturdy, dangerous spine that is sharp and poisonous.

 Fish caught by the Inuit, indigenous people of northern Canada and parts of Greenland and Alaska, **freeze** naturally and are preserved for months.

 Ice diving is scuba diving in ice-covered waters. It can be dangerous and requires special equipment and techniques.

 The **narwhal** has a long horn that is the source of the legend of the unicorn. Its horn is actually a fully developed tooth that can grow to be 9 feet (2.7 m) long.

 The **polar bear**, one of the largest carnivores alive today, has extremely thick fur that protects it from the cold and helps it float when it swims.

 The **white-beaked dolphin** is a typical dolphin of the waters off Iceland. It lives in groups that move quickly, performing acrobatic stunts and leaps, and loves to follow boats.

 Beached wood and the remains of hunted seals were used to build **kayaks** hundreds of years ago.

Orcas in the eastern Atlantic migrate in the winter, along with other large cetaceans, to reach the waters between Iceland and Norway, following the herring and mackerel.

 The **Vikings of Greenland** settled on the island approximately 1,000 years ago during what some historians believe was a warm period. They died out a few hundred years later, perhaps when the climate began to cool.

 The **walrus** is related to seals, but they are larger, have tusks, and have wrinkly skin instead of true fur.

 Sólfar, or the **Sun Voyager**, is the monument that commemorates the founding of Reykjavík, in Iceland. It resembles a Viking ship and symbolizes hope for the future.

 The **polar cod** reproduces only once in its life, and the females lay on average about 12,000 eggs, compared to the 9 million laid by Arctic cod females.

 The **Greenland shark** can grow up to 23 feet (7 m) long and is dangerous to humans. Its cold-water existence allows it to live up to 400 years.

 In the northwest British islands, the **coracle** has been used for thousands of years. This small boat is made from willow and animal skin, and steered with a paddle.

 The Inuit of Greenland warded off attacks using **tupilaqs**, small monsters carved into bone, which they lowered into the sea in order to find and kill their enemies.

 Four cement legs that are 1,213 feet (370 m) high support the **Troll A** natural gas platform, which in 1996 was towed for 124 miles (200 km) and then positioned on the ocean floor.

 The **cusk** is a cold-water fish with an elongated body, shaped almost like an eel. It lives on rocky or pebbly ocean floors between 656 to 1,640 feet (200 and 500 m) deep.

 Herring is fished in great quantities and is commonly preserved by **smoking**, a process carried out for centuries in the northern countries.

 The **northern fulmar** is a seabird similar to the seagull. It is better suited to flight than to life on land because of its small feet, which make walking awkward.

 The fiery Viking **Erik the Red** was forced to seek a new home after being exiled from Iceland and founded Greenland's first colonies.

 The **ocean quahog** is perhaps the most long-lived known animal. One lived to be 507 years old.

 In Norway, a **tunnel** is being designed through which ships will to ensure the safety of those traveling along a stretch of coast struck by dozens of storms every year.

 The **white-tailed eagle** is the fourth-largest eagle in the world. Its wingspan can reach nearly 8 feet (2.5 m).

The Arctic and Antarctica

The Arctic Sea is very cold and unwelcoming, but it is surrounded by habitable lands. It is also the only ocean in the world that can be walked on, since much of it is covered with ice. Thousands of years ago, the first "ice people" ventured across this ice desert.

Over time, the Arctic has seen numerous cultures develop. They have been very different from each other, but they are always tied to the resources of the sea, since the climate isn't suitable for farming. Arctic populations became sea hunters and fishermen thanks to tools such as harpoons made from walrus bone and kayaks made from waterproof sealskin.

The Arctic was later the destination for a very different migration, which began approximately 1,000 years ago, by European seafarers searching for new lands. They were followed by the exploration of regions of Russia and Canada to the north, in a quest for commercial routes between the Atlantic and Pacific oceans.

Ancient peoples believed that there was an unknown land at the bottom of the planet that balanced the weight of the lands located to the north. In fact, old globes sometimes show a continent similar to Antarctica.

The exploration of the Arctic Sea is important for maritime transport and the extraction of gases and minerals. However, the Arctic Ocean is fragile, since this frozen sea is melting. In fact, today this region has become a barometer for measuring the world's health, and it is carefully monitored by research ships and by futuristic instruments like HydroBats.

The Antarctic region is a continent surrounded by thousands of miles of ocean thrashed by storms, with few arid islands making it habitable for humans. The seas are also difficult to sail, and kept navigators at a distance until 200 years ago.

About 10,000 years ago, human beings who arrived farther south than anyone had before stopped at the large isle of Tierra del Fuego, close to South America. Only 700 years ago, the Polynesians attempted to settle on the sub-Antarctic islands of Auckland, but they soon had to abandon them.

It was only in 1820 that the Russian sailing ships *Vostok* and *Mirny* pushed far enough south to see Antarctica, and another 75 years would pass before anyone went ashore. Very few islands in these remote waters were inhabited, and they usually served as bases for whaling ships and exploration vessels such as the *Fram*, used by Roald Amundsen, who reached the South Pole in 1911.

Today the Antarctic Ocean is crossed by fishing boats, sailboats, cruise ships, and supply ships heading toward the few inhabited places. Some of these places are very important scientific bases where researchers live and explore the region using micro-submarines.

The Arctic region is located north of the polar Arctic Circle and is made up of a large sea surface, the glacial Arctic Ocean, surrounded by various portions of land belonging to different continents. During the winter, these lands are almost completely joined by a sheet of ice that forms over the ocean, called the ice pack.

In recent years, global warming has been impacting all Arctic life and ecology. The increase in temperature blocks the formation of ice sheets that are increasingly thin and last for increasingly short periods, endangering the life of polar bears and seals and creating problems for Eskimos and others who live on land.

40°F (4°C)

Cold temperatures are not a problem for life in this frozen habitat. Arctic lands and seas are the ideal environment for many life forms, including species of great biological interest, such as the polar bear, the narwhal, the beluga, the walrus, the bearded seal, and the hooded seal.

The Arctic is full of great rivers. Some, like the Ob, run through Siberia and empty into the Kara Sea, carrying large quantities of fresh water into the Arctic oceans. The Siberian sturgeon swims back up the Ob to reproduce.

The Antarctic is a true continent covered by ice for much of the year. It is surrounded by the Antarctic Circumpolar Current, the strongest current in the world, which runs constantly around the Antarctic at depths of up to 9,850 feet (3,000 m), carrying 4,767,000 cubic feet (135,000 m³) of water per second.

Antarctic Circumpolar Current

The Ross Sea is a large bay that extends into the Antarctic and is covered by the largest floating glacial platform in the world, which extends for more than 280 miles (450 km) from the coast. Emperor penguins and seals live on this platform, moving between the sea and the ice.

The Amundsen-Scott Scientific Research Station is located on a moving glacier near the South Pole at about 9,300 feet (2,835 m) above sea level. It was built in 1956. Scientists there study astronomy, weather, and earth science, among other things.

The life of all large Antarctic animals, such as penguins and other birds, seals, and cetaceans, depends on a single organism, a small crustacean that is up to 3 inches (7 cm) long, known as krill. Millions and millions of krill exist in seas around the Antarctic continent and the Antarctic Circle.

It was once thought that the waters of the Antarctic were too cold to allow fish to live. Research today has shown that these waters are populated by more than 260 species, for the most part belonging to a very specialized group that began evolving 30 million years ago.

Arctic Ocean

Link to download Arctic Ocean map:
www.blackdogandleventhal.com/
MapsoftheWorldsOceans/Arctic_Ocean

East Siberian Sea

New Siberian Islands

Dmitry Laptev Strait

SIBERIA

Laptev Sea

Ocean

Ridge

NORTH POLE

North Pole

Garkel Ridge

Nansen wetlands

Northern Land

PACIFIC OCEAN

North America

ARCTIC OCEAN

Asia

ATLANTIC OCEAN

Europe

Africa

Dikson

Tazovskiy

Franz Josef Land

Novaya Zemlya Trench

Kara Sea

Tambey

Gulf of Ob

578

578

Svalbard Island

NORGE

Nordaustlandet

Novaya Zemlya

39°F (4°C)

Longyearbyen

Edgeøya

Stolbovoy

Spitsbergen

Barents Sea

RUSSIA

The Arctic Circle

EUROPE

131

Arctic Circle

 Known as the **northern sea nettle**, the *Chrysaora melanaster* jellyfish has an umbrella-like body up to 2 feet (60 cm) wide, with tentacles that can be up to 10 feet (3 m) long.

 The **bearded seal** is named for its long, dense whiskers (called vibrissae) that sprout from its snout. Young bearded seals plunge into the water just a few hours after birth.

 The **glaucous gull** is a large seabird with a wingspan of more than 5 feet (1.5 m). It nests along the most northern coasts.

 Tin lifeboats with motors and medical equipment such as hyperbaric chambers are used today to rescue shipwrecks in the icy Arctic waters.

Inuit fishermen have long used **protective goggles** made out of wood or bone to keep them from being blinded by the reflection of the sun on the sea and on the ice.

 The **snow crab** is a colorful crab that can be more than 6 inches (15 cm) wide and can weigh up to 3 pounds (1.3 kg).

 The **pink salmon** swims up the rivers of Alaska and Siberia to reproduce. During this journey the males change their appearance, developing darker coloring and a large hump.

 In the Arctic region there are many different myths about **Sedna**, goddess of the sea and sea creatures. They all claim that the first seals, walruses, and whales emerged from her fingers.

 In the nineteenth century, it was thought by some that a **temperate open ocean** lay beyond the polar ice. This mistaken idea cost the lives of many explorers, including Franklin.

 The **black guillemot** is a black seabird that uses its wings to swim underwater and hunt fish.

 Frankenstein's monster from Mary Shelley's 1818 book, *Frankenstein*, flees to the Arctic Circle.

 The **glacier lantern fish** can live at depths of 4,600 feet (1,400 m) and migrate for long distances underwater, rising up at night to feed on plankton.

For centuries the people of the Arctic have hunted sea mammals with **harpoons**, once made of wood and bone, now made of metal (sometimes using old golf clubs!).

 The **hovercraft** is a vehicle used to move around the Arctic regions. It can move without stopping on water, ice, snow, and shores that are not too steep.

 The **ulu** is an ancient crescent-shaped chopping knife, typical of the Arctic. It is used for cleaning fish and other prey and also for cutting blocks of ice to make igloos.

 Now that the icy ocean in which **polar bears** live is gradually melting, some people think they will be able to survive only in captivity, such as in zoos.

 Zarya is the name of two important Russian research ships. One was used for polar exploration in the early twentieth century, and the other was used 50 years later to study the magnetic field of the earth.

 Inuits used the **umiak**, a cargo boat made of wood and sealskin, for transporting people and furnishings. Women did the rowing, singing to produce a rhythm.

 In 1773, Russian merchant Ivan Lyakhov claims to have discovered the island of **Kotelny** while following reindeer tracks on the oversea ice. However, he found a kettle left there by an earlier explorer.

 In 1845, British naval officer Sir John Franklin led an exhibition to explore the Arctic and the Northwest Passage aboard two ships, the **Erebus** and the **Terror**. The ships became trapped by ice, and his crew eventually died. The *Erebus* and the *Terror* now lie at the bottom of Queen Maud Gulf.

 The **hooded seal**, named for the male's snout, has a sort of elastic balloon on its head that it can inflate with air.

The monster **Qalupalik** is a bit like the wolf in fairy tales. It punishes disobedient children, but instead of eating them, it takes them to live with him in the dark depths of the sea.

 The **Siberian sturgeon** can grow to more than 6 feet (2 m) long and can weigh more than 440 pounds (200 kg).

 The icy Baffin Bay is a popular destination for **stand-up paddle-boarding**. Enthusiasts can move between blocks of ice, using ordinary planks of wood, even for long routes.

 A gigantic **gas platform** operates amid the ice of the Kara Sea, where rich natural gas deposits are being discovered thousands of feet deep.

 The **narwhal** is known for the long ivory tooth coming out of its mouth that looks like a horn. Usually males only have a horn, but sometimes females have them as well.

 The Russian steamship **Yermak** was the first polar icebreaker in the world. It was launched in 1898, and was even used in World War II.

 Flatfish is a species of fish that can camouflage themselves on the sandy bottoms of the ocean. They usually have long snouts and a dip near their eyes.

 The **king eider** is an Arctic duck with a black body, white chest, and multicolored head with noticeable frontal bump.

 The **Atlantic salmon** can grow up to 30 inches (76 cm) long and weigh up to 12 pounds (5.5 kg).

 Living in isolation for centuries in the Hudson Bay, the Sadlermiut people fished with harpoons on board inflatable rafts made of **inflated walrus skins**.

 The main ingredient of **Suaasat**, the typical soup of Greenland, is often made from the meat of seals, whales, seabirds, or reindeer. It is a hearty meal suited to the cold climate.

 The **ringed seal**, the most common seal in the Arctic, is also the smallest. It spends its life among the ice, choosing areas with cracks where it can plunge into the water.

 Gray whales are true acrobats. They perform spectacular leaps outside the water, with most of their bodies exposed before falling back down into the ocean.

 The official conquest of the North Pole was carried out by Norwegian explorer Roald Amundsen on board the **Norge zeppelin** that flew over the Arctic in 1926.

 The **long-tailed jaeger**, named for its very long tail feathers, looks like a white and black gull, but it is actually a true Arctic predator.

 In the Greenlandic Inuit religion, *tupilaq* were monsters that were made out of **walrus bone** by a shaman, a kind of priest, that were sent out in the ocean to kill an enemy.

 The first **canned foods** were not very safe to eat and are in part responsible for the tragedies experienced by Arctic navigators who were often poisoned by rotten and/or polluted food.

 The **USS Skate** was the first submarine to emerge from the ice precisely at the North Pole in 1959.

 The **beluga** or white whale is one of the most fascinating cetaceans. Unlike other whales, it has an actual neck that allows it to move its head from side to side.

The Arctic:
Polar Bear Kingdom

The Arctic is a continent of ice. It can only exist if winter temperatures allow the seawater to freeze.

The waters of the Arctic are inhabited by different species of seals, fur seals, and walruses. These mammals were heavily hunted until the middle twentieth century but are now a protected wildlife.

Antarctic Ocean

Indio-Atlantic Basin

Enderby Land

South Georgia Island

South Sandwich Trench

South Sandwich Islands

South Orkney Islands

South Shetland Islands

Cape Ann

Cooperation Sea

Cape Boothy

Riiser-Larsen Ice Shelf

Weddell Plain

70th Parallel

Brunt Ice Shelf

Queen Maud Land

Cape Darnley

Bellingshausen Plain

Graham Land

Larsen Ice Shelf

Weddell Sea

Amery Ice Shelf

Antarctic Peninsula

Filchner-Ronne Ice Shelf

80th Parallel

A N T A R C T I C A

Davis Sea

Bellingshausen Sea

Ellsworth Land

WESTERN ANTARCTICA

+ South Pole

EASTERN ANTARCTICA

Shackleton Ice Shelf

Getz Ice Shelf

Marie Byrd Land

Ross Ice Shelf

Wilkes Land

80th Parallel

Amundsen Sea

Victoria Land

Ross Sea

Cape North

Pacific-Antarctic Ridge

Scott Island

Balleny Island

D'Urville Sea

South Indian Basin

ATLANTIC OCEAN

Africa

ANTARCTIC OCEAN

INDIAN OCEAN

South America

ANTARCTICA

Australia

PACIFIC OCEAN

41°F (5°C)

Link to download Antarctic Ocean maps:
www.blackdogandleventhal.com/MapsoftheWorldsOceans/Antarctic_Ocean

Ross Sea

Amundsen Coast

Gould Coast

Dufrek Coast

Siple Coast

Richards Inlet

Shackleton Coast

Crary Bank

Shirase Coast

Ross Ice Shelf

Penny Point

Inlet of Barne

Cape Kerr

Roosevelt Island

Saunders Coast

Edward VII Peninsula

Sulzberger Bay

Moore Bay

Hillary Coast

Ross Island

Byrd Canyon

Bay of Whales

Glomar Challenger Basin

Marie Byrd Land

Ross Bank

McMurdo Channel

Nordenskjöld Basin

Victoria Land

Scott Coast

Pennell Bank

Ross Sea

Drygalsky Basin

Coulman Island

Borchgrevink Coast

28°F (2°C)

Scott Canyon

Cape Hallett
Moubray Bay

Hillary Canyon

Pennell Bay

Cape Adare

Robertson Bay

Cape North

ATLANTIC OCEAN

INDIAN OCEAN

ANTARCTICA

PACIFIC OCEAN

ROSS SEA

Antarctic Ocean

 Rusted whaling ships can be found around the Antarctic from times when hunting cetaceans was not regulated. Today it is forbidden in the waters, but not all nations have recognized the ban.

 If the Antarctic polar icecap were to melt, the **sea level** would rise 213 feet (65 m), nearly as high as the Statue of Liberty.

 Remote-control **autonomous underwater vehicles (AUVs)** are submarines that can descend 16,400 feet (5,000 m) and can navigate while submerged for more than 62 miles (100 km) beneath thick layers of ice.

 In 2011, two large **submerged volcanoes** were discovered a few feet from the surface near the island of South Georgia.

 Some icebergs in the Antarctic Ocean are so large that they change the shape of the **clouds** above them, creating long, rippled trails behind them.

 Scientists, video operators, and wealthy enthusiasts can explore the waters with **ultralight submarines** adapted to the extreme conditions of the Antarctic.

 The **Gray's beaked whale** is a rare species of toothed cetacean that lives in the Ross Sea. It is about 20 feet (6 m) long and weighs up to 2,400 pounds (1,090 kg).

 Ottoman admiral and cartographer Piri Reis created a **world map** in 1513. Only a third of the map still exists, and it is unclear whether or not the full map included Antarctica.

 In 1915, the *Endurance,* the ship of British explorer Ernest Shackleton, became frozen into an ice floe and started sinking. Shackleton led his men to safety on **lifeboats,** crossing in open sea for five days to reach Elephant Island.

 About 150 years ago, the writer Jules Verne imagined an adventure in the Antarctic seas, on board the **Nautilus** submarine led by Captain Nemo in his famous book *Twenty Thousand Leagues Under the Sea.*

 In order to dive into the icy waters of the Antarctic, divers must have many certifications, wear a special **heated wetsuit**, and be perfectly healthy, trained, and fit.

 The distant Antarctic waters have inspired various writers such as H. P. Lovecraft, who set the fictional lost city of **R'lyeh** from his short stories here.

 The so-called **Arctic cod** can even live in waters with temperatures below zero degrees because its body contains special antifreeze material.

 The **Fram** was a ship built and designed in 1892 to withstand being crushed on the ice. Norwegian explorer Roald Amundsen sailed the *Fram* to reach the South Pole in 1911.

 The **Pagothenia**, or bald notothen, like the North Sea herring, is very common in the Antarctic oceans, where it can live in waters close to freezing temperatures, thanks to natural antifreeze in its blood.

 The **SY Aurora** was a steam yacht built in 1876 that rescued trapped ships in the Arctic and Antarctic seas until 1917.

 Abnormal waves as large as 65 feet (20 m) high can form unexpectedly in the open ocean. They are rarely observed by human eyes.

 Rare **racing sailboats** brave the Antarctic Ocean, venturing for thousands of miles in the deserted waters as far as Cape Horn, then turning back northward in the Atlantic Ocean.

 Orcas can be found in every ocean, but they are most abundant in the southern seas, where about 25,000 of them live today.

 At the Scott Base in Antarctica, there is a sculpture honoring the **Maori seamen** who ventured into the Antarctic Ocean centuries ago. It is believed that they landed on the frozen Ross Ice Shelf.

Ross Sea

 The naval officer **James Clark Ross** explored the oceans of both poles, but he is best known for exploring the Antarctic in 1841.

 Before it moved to the bottom of the planet 25 million years ago, Antarctica was covered with **forests**.

 Luxury cruises that carry fortunate passengers around the world also reach the southernmost waters of the globe, beyond 78 degrees south latitude.

 In the springtime, the **Antarctic petrel** flies for hundreds of miles toward the mountains of the Antarctic interior, where it nests.

 Icebergs also get trapped by the ice of the Ross Sea, creating **flat mountains** in an immense plane.

 The **south polar skua** is a predator bird that threatens penguin chicks and eggs and steals other birds' prey.

 The **Adélie penguin** is the smallest penguin in the Ross Sea, but the most plentiful, numbering about 6 million. It is 18 to 29 inches (45 to 75 cm) tall.

 The **leopard seal** is a nightmare for penguins and small seals. Along with the orca, it is the greatest predator of the Arctic seas and has sharp, cutting teeth.

 Modern **polar pyramid tents** are based on the design of British explorer Robert Falcon Scott's tent from the early 1900s.

 The **Pleuragramma**, or Antarctic silverfish, is one of the most abundant fish in Antarctic waters. It usually lays its eggs by the millions, amid ice that is forming.

 The **ANDRILL** boring device drills down through more than 3,900 feet (1,200 m) of ice and rock to reach the bottom of the Ross Sea, in search of precious information about our planet's past.

 The Antarctic **vase sponge**, or glass sponge, is a gigantic silica sponge, up to 6.5 feet (2 m) that covers vast areas of the Antarctic ocean floor.

 The **Antarctic cod**, also known as the Antarctic toothfish, is not a true cod, but it is a giant among the fish of the Antarctic, and can be up to 5.5 feet (1.7 m) long and weigh up to 300 pounds (135 kg).

 The **Arnoux's beaked whale** is a large cetacean that is almost 33 feet (10 m) long. It has a long beak and a longer lower jaw, making its front teeth always visible.

 Getting supplies to polar stations can often be difficult. It sometimes requires using icebreaking ships to **move icebergs** that block the few navigable routes.

 The **Weddell seal** lives along all the coasts of the Antarctic continent, preferring to stay on ice rather than land.

 The **Antarctic minke whale** is very similar to other minke whales that live in the North Atlantic and the northern Pacific, but it is slightly smaller in size.

 British ocean advocate Lewis Pugh swam in Antarctic waters in only a **bathing suit** in 2015 to gain attention for creating a protected marine region in the Ross Sea.

 The **giant squid** can be found in the cold Antarctic waters. It can grow up to 43 feet (13 m) long and can weigh up to 600 pounds (272 kg).

 The **ARTEMIS** is an autonomous underwater robot that is immersed into the Ross Sea to gather scientific data and collect water samples.

 The **emperor penguin** is the largest of all penguins and can be more than 4 feet (1.25 m) tall. Every winter, these penguins walk up to 75 miles (120 km) to reach their breeding grounds.

 In 1910, British explorer Robert Falcon Scott led an expedition on the **Terra Nova** to reach the South Pole. They did reach their goal, but Norwegian explorer Roald Amundsen got here 34 days earlier.

 An ice shelf is a thick platform of ice that forms where a glacier meets the ocean. The Ross Ice Shelf is the largest ice shelf in Antarctica, and is a destination for extreme tourism like **kayak cruising** today.

Antarctica:
Where the Penguins March

Krill are similar to shrimp. They are small but the Antarctic is full of them and they are necessary to the circle of life on and around the continent.

People used to think the sea floor of the Antarctic was almost lifeless. We now know that it is actually inhabited by a large number of animals and plants with unique and extraordinary features.

Index